Maurice Burton

A zoo at home

J M Dent & Sons Ltd

London   Melbourne   Toronto

First Published 1979
© Maurice Burton 1979

Phototypeset in 11/12½ pt VIP Baskerville
by Trident Graphics Limited, Reigate, Surrey

Printed in Great Britain by
Billing & Sons Ltd, Guildford, London & Worcester for
J. M. Dent & Sons Ltd
Aldine House, Welbeck Street, London

ISBN 0 460 04383 8

Contents

List of plates

The common jay, when anting, brings its wings forward and arches them as if forming a canopy

A common rat climbing a metal rod

A pair of mallard resting

A corner of the garden at Weston House

Sanctuary 1

The British Army, of which I was an insignificant member, and the French Army, reeling under the final German offensive that began on 21 March 1918, had been driven back. To save our guns my battery had been pulling back day after day in a monotonous and gruelling retreat, losing men as prisoners of war, killed or wounded on the way. At last we reached a position near a French village where attackers and defenders, from sheer exhaustion on both sides, settled down to dig in and hope. It was an interval of calm between the disastrous retreat and the movement forward which ended the war.

It so happened that at this final point of our retreat we found ourselves occupying a line of trenches that had been held by the French army in 1914, in stemming the original German advance on Paris. It was historic ground, therefore, a place symbolic of change in the fortunes of nations. It was historic for me, also, for there occurred an event that was to change my whole life. On a beautiful sunny afternoon in early May, I sat and watched the activities of some ants which had their nest in the crumbling parapet of an ancient trench. Oblivious to the sounds of war or the fate of nations, engrossed and enthralled, I watched the ants for two hours or more. They were equally oblivious to outside events: they were going about their daily business.

It so happened that among the personnel of our battery was one by the name of Hooker, but known to his comrades as Daddy, because of his portly build and his paternal attitude to all the men. Bursting with enthusiasm about my discovery I felt impelled to pour out my observations to the first man I met – who happened by some heaven-sent chance to be Daddy Hooker. He listened patiently, then began to talk about matters of natural history; he revealed that in civilian life he was a lecturer in the subject. Before enlisting, in May 1916, I had planned to read history at university. Now, inspired first by the ants and then by Daddy Hooker, I turned to natural history. I wrote home immediately asking for books on zoology. Two books duly arrived and I carried them with me throughout the campaign. When the armistice came, on 11 November 1918, I knew the contents of those two books almost by heart.

We were on our way to the Army of the Rhine, after the cessation of hostilities, and had reached Namur, in Belgium, when I learned that I was going home on leave. Even greater news was to follow. Those going on leave were told that if they had jobs to go to they need not come back. I arrived home at midnight on the Boxing Day of 1918 and the following morning I set out for King's College, London, there to enrol as a student. If by enrolling as a student I could legitimately remain at home, to await demobilization, then I would leave no stone unturned to achieve this end. I had willingly joined the army but I would as willingly leave it once the fighting was over. The formalities took some days but finally were completed and I joined the Zoology Department. By some miracle, which I have never fully comprehended, I passed the first examination the following summer successfully, so qualifying to go forward to a degree course.

In 1921 I took an honours degree in zoology. For this I needed to take a special subject. I chose sponges, or rather sponges were chosen for me. The decision to do so had had to be taken in the summer of 1919, as soon as the results of the first exam were known. This was a bare nine months after the end of the Great War and only a few months less since my final release from the British army.

Like so many of my contemporaries, in that period of mud and turmoil, I had lost any capacity I may have had for planning a future. The venerable Professor Arthur Dendy, my future mentor, suggested I should study sponges, the *Porifera*, to give them their scientific name. He pointed out that he was a leading authority on them and I should therefore have the benefit of his personal supervision and guidance. So I readily assented.

If I had hoped for more ants, what I got was sponges! Truth to tell, I was in such a mood of mental drift and disorientation, as a result of war experiences, that I would have agreed to almost any line of study he might have suggested. I was to learn later that most zoologists treated this obscure group of organisms with something approaching amused contempt. In the same year as I had graduated A. P. Herbert wrote in *Punch*:

> *The sponge is not, as you suppose,*
> *A funny kind of weed;*

He lives below the deep blue sea,
An animal like you and me,
Though not so good a breed.

I also discovered that Professor G. C. J. Vosmaer, of Leiden, another leading authority, had described sponges as the pariahs of science. Sir Gavin de Beer was emphatic that sponges were not even animals. This was many years later, when Sir Gavin was director of the British Museum (Natural History), better known as the Natural History Museum, in which my professional life was spent. We chanced to meet one morning on the steps leading to the main entrance.

'Burton,' he said, without preamble, 'sponges aren't animals.'

With this he bustled up the steps in front of me and, as if to make sure I had heard him aright, he flung over his shoulder: 'Sponges aren't animals.'

When I asked him what they were, he said: 'I don't know, but they are not animals.'

Sponges, I found, are difficult to understand, even more difficult to study. They are quite unlike any other kind of animal. They are enigmatic, baffling. Yet, like sea-bathing in winter, their study is awful to contemplate but exhilarating once you are fully immersed. After graduating I became so immersed in my study of sponges that I devoted my whole time to them. I flung myself wholeheartedly into their study and continued to do so when I took up an appointment at the Natural History Museum. Then I began to have nightmares.

The climax came one night when I dreamt I was dissecting a human corpse. When I had reduced it to the bone, the skeleton sat up on the bench and started talking to me. I awoke in a sweat, horrified, and I realized the truth of another of Professor Vosmaer's remarks: 'There is a curious connection between the study of sponges and insanity.'

What was needed, I realized, was a counter-balance to offset my preoccupation with this specialist study – so I took up field natural history as a hobby. This gave me exercise, fresh air and an absorbing interest in the widest possible subject, the whole living world. My subsequent marriage, to Margaret Rosalie Maclean, also helped to give me a balanced outlook on life, and as soon as my children were able to participate in them, the field studies took on a new aspect.

Jane, especially, born on 19 August 1933, introduced an element of teamwork for, from an early age, she began, on her own initiative, keeping pets. She has, as it is called, a way with animals and is a born naturalist. The team was enlarged when Richard was born, in 1935, and further extended when Robert was born, in 1941.

When World War II ended we were living in Twickenham, on the outskirts of London, and it was not long before the family was making regular trips into the countryside, on natural history rambles. Since I no longer owned a car the journeys between home and the open country, by bus and train, were arduous, so in 1952 we decided to move into the country, to have the wild life 'on our doorstep'.

In that year we moved to Horsley, in Surrey, then a rural village in the Green Belt. The house was an ordinary suburban dwelling and it stood in a third of an acre of ground, a quarter of which was wild and contained two well-grown oaks, a hornbeam and a plum tree. Clearly this was ground that could not be cultivated. Soon after we had settled in, Robert found a hedgehog wandering along the shady hedge-bottom on the western boundary of the garden at midday in the broiling sun. This was unusual, for a wholly nocturnal animal to be out and about at noon.

We had seen hedgehogs before but we had never possessed one. Robert was highly excited and without a pause for thought we decided to erect a wire-netting fence to enclose the wild part of the garden. There we would keep the hedgehog so as to have it always under observation. All set to work and by evening the fence was finished. Care had been taken to secure the netting, especially at ground level, so that the hedgehog could not escape. Needless to say, on the following morning there was no sign of it. It had found the one weak spot in our fence, had prised up the netting sufficiently to crawl under it and was gone.

We were naturally disappointed, but from that disappointment sprang the idea of converting this wild corner of the garden into a sanctuary. If successful we would then, quite literally, have the wildlife at our doorstep. We even drew a plan of what the sanctuary would look like – sixty feet by fifty feet with a four-foot wire-netting fence enclosing it and a small wooden gate to give access. The brambles and scrub would be allowed to spread and we would plant shrubs to bear berries in autumn and winter. Nesting-boxes would be put up on the trees and we would put logs and branch litter on the ground to provide shelter for small mammals. There was already a small pond there and this we planned to extend. We did put up the fence and we engaged two workmen to enlarge the pond. Unfortunately it never held water. Meanwhile we were discovering plenty of small wild life in the garden itself, including shrews, fieldmice, voles and rats. The latter were not too welcome, although they had their compensations.

There was one morning when we were clearing a pile of brushwood. All went smoothly until we reached the bottom layer. When this was lifted fieldmice leapt in all directions. It was like disturbing a colony of tiny kangaroos. As they bounded this way and that there seemed to be scores of them. In fact, there were fewer than twenty. One jumped from the ground into a shrub, a standing jump of four feet – amazing for an animal little bigger than a house mouse. I once surprised a grey squirrel under one of the oaks. It leapt ten feet up from a squatting position, clutched the outer twigs and swarmed up into the tree. It seems incredible, yet I measured it after the twigs had ceased swaying. The unusual athletic feats of some animals have held a special interest for me since that day.

We did nothing more about rehabilitating the pond because we soon had our hands full with other things. Somebody learned we might be prepared to take over a pair of tawny owls. They had been abandoned by the parents. Then someone else asked us to look after a genet, a cat-like animal related to the mongoose. About the same time I saw an advertisement in a newspaper for a fox cub for sale. A few years previously I had read all I could find about the red fox and the results had been far from satisfying. I had found plenty of stories about the hunting field but although most authors described the fox as very intelligent few gave any reason for saying so. So I bought the cub to see for myself, if possible, what this intelligence amounted to.

From there it snowballed and in the five years that followed we had nearly one hundred and fifty mammals and birds, representing over fifty species, as more or less permanent residents in the garden or in the house. During the same period nearly five hundred others needing care and attention were brought to us. Most of these were set free as soon as they could fend for themselves. We kept only those that were unlikely to survive in the wild.

The main burden fell on Jane, who was responsible for the hospitalization, and she would spend days on end with saw, hammer, nails, staples and wire-netting, building aviaries and pens. Bad weather did not deter her. I can still see her now in her old raincoat and sou'wester, in pouring rain, sawing and hammering to her heart's content. Her enthusiasm has never flagged throughout the twenty-five years that have elapsed since we first embarked on a venture that was to grow far beyond our original dreams. Indeed, it was her drive that kept it going at an ever-increasing pace, although to say this is not to detract from the enthusiasm shown by other members of the family.

One hot day, for example, when we had not gone beyond the sanctuary idea, I said: 'If we are to encourage animals, especially birds, into the sanctuary we must provide the right food; and we had better start storing it now.'

'What kind of food?' asked Robert.

'Berries and nuts, especially rose hips and haws,' I replied.

An hour later, I saw Robert come bustling purposively through the front gate, sweating profusely, carrying a large basket filled with hips, his face as scarlet as the berries he had been gathering. He must have moved with frantic energy to have collected so many in so short a space of time. Robert has the faculty of immediate enthusiasm, carried through wholeheartedly. But if the enthusiasm of the others is less obvious it has been none the less real and sustained.

Within a year the garden, which had looked ample enough when we started, was steadily filling since we always made their enclosures as large as possible to give each animal or pair sufficient space. The birds, for example, were given enough room to spread their wings and fly from one perch to another. Also, each aviary and each pen was furnished with branches or other vegetation to give as near the natural habitat for the species as could be contrived.

The advantage for me was that no matter which way I looked, indoors or out of doors, there was a constant panorama of animal behaviour. I felt that at last I was truly becoming a zoologist; and there were also opportunities for more intensive studies. The advantages for Jane were that she was finding an outlet for her natural abilities as an animal keeper as well as abundant opportunities for photographing animals, and photography was fast becoming her métier.

The disadvantage came when we decided we must find another house, with a larger garden – which meant, of course, selling the cottage we were living in. Plenty of people came to inspect it with a view to purchase. None wanted to buy. We soon began to recognize the symptoms. As soon as a prospective buyer came through the gate and looked around the garden, jam-packed with aviaries and pens, we would see that person's expression change for the worse. It grew darker still as the tour moved to the house. There was an animal of some sort in every room. Each visit ended with what was to become a drearily monotonous remark: 'We will let you know.'

One day my wife, taking a prospective buyer around the house, showed off the bathroom. 'This is the airing cupboard,' she said brightly, and opened the cupboard door. On a shelf, among the neatly folded household linen, was a nest of young polecats Jane was hand-rearing.

Eventually we sold the cottage, at a loss, and moved into Weston House, in Albury, five miles south of Horsley. To get a large garden it followed we had to buy a large house, at a large price. Fortunately, both house and garden proved to be all we could ask for. The garden itself, more especially, was ideal. It occupied five acres and was well established, filled with well-grown trees and shrubberies as well as a kitchen garden, orchard, flower-beds and lawns. Owls nested in one of the trees, frogs spawned in one of the ponds, jackdaws nested in the chimneys of the house, a pair of weasels had their home in the rockery. Shrews, fieldmice, voles and rats could be seen any day. Bats flew over the garden of an evening; and birds nested everywhere in the shrubs and hedges. To crown all this, the garden was surrounded on three sides by fields. It was a naturalist's paradise, as much for the wild flowers as for the animals.

All the aviaries and pens in the previous garden had been dismantled and transferred to the new garden, along with their occupants. Here, however, it was possible to site the aviaries and the pens so that they were virtually hidden among the shrubberies. It was also possible to build more of them and still not disfigure the garden, and our accommodation for keeping animals was soon doubled.

Nature, we are told, abhors a vacuum. In that case, Jane is a true child of Nature, for if there is a vacant space somewhere she is not happy until it holds an animal of some sort. In addition to the two ponds, which were here when we came, a third was built and all three had to have fish in them. A fourth, much smaller, was built for pond tortoises and terrapins. Two donkeys were acquired. They grazed the largest of the lawns for a while but were later pastured in an adjoining field. A nanny goat, bought while still a kid, was tethered on one of the smaller lawns.

A chicken house was bought and another lawn became a poultry run. A piece of ground in the shade of a large ornamental *Prunus* tree, under which nothing but sparse weeds would grow, became the site of a run for bantams. Mallard and Muscovy ducks somehow found their way into the garden. How Jane acquired them all we seldom asked. Sufficient was it that she could not refuse any offer of any kind of animal; and I rapidly caught the fever, too. After all, we had the space and the willingness to look after them. Even the gaily coloured gypsy wagon, Jane's prized possession that added such a touch of gaiety to the garden, was used to house injured birds or young animals brought to us for hospitalization. None was ever turned away. An injured swan and two starving bitterns were among the more spectacular of the many that found salvation in the garden of Weston House. All three were given their freedom once they had recovered.

Within a year of our taking it over, these five acres of ground were more like a wildlife park than a country estate as the Muscovies, now numbering fifty by natural processes, mallard, bantams and domestic fowl, wandered freely everywhere. The cost of fencing to keep them from the kitchen garden and the flower-beds was heavy.

Inside the house, a large room on the ground floor, opening to the garden, became Jane's studio. It was crammed literally from floor to ceiling with cages while aquaria and vivaria were stacked around the room.

Four years after we moved to Weston House, Jane married. Her husband, Kim Taylor, is a zoologist and soon after their marriage they went abroad, to Ghana, for three months. It was necessary before their departure to scale down the number of animal occupants, simply because I had not the time to deal with everything single-handed.

As soon as Jane returned, however, the population of our 'zoo' soared again. Then Kim was sent to Kenya for six months. Again, numbers had to be scaled down. This became a regular pattern for, in the years that followed, the Taylors were often abroad, with long spells at home in between. After six months at home from Kenya, they went to Malaya for six months. Then another six months home and off they went to Barbados for a year, followed by a spell at home and then another nine months in Kenya. The number of animals in our zoo fell and rose spectacularly as the Taylors alternately went abroad and returned home. At each scaling down, when new homes were found for many of the inmates of the zoo, the 'old-stagers' – the owls, genet, foxes and others – that figure prominently in the chapters that follow, still remained with us, living in most instances to a ripe old age.

We started with the idea of giving sanctuary to wild life on a very small scale. Many other people up and down the country have done, and are still doing, the same thing. Side by side there developed the humanitarian motive of rescuing injured and abandoned animals – again in a humble way, and achieving no more than several people or groups of people have done, in Britain and elsewhere in many parts of the world.

Robert, who was only ten years of age when we started our sanctuary, went on to become a zoologist and is now an author who loses no opportunity to visit remote places. Richard, his elder by six years, like Robert, read zoology at Cambridge University, but later switched to physiology and is now a lecturer in this subject at Glasgow University. Both must owe something to their daily childhood contact with live animals. The more spectacular result is seen, however, in Jane, who has been described as 'the leading animal photographer in the country'. Her colour pictures of animals may be seen in scores of books and magazines and in the books she herself has written. All three must have been influenced by the quiet, though not always unspectacular research, that continued daily over a period of twenty-five years, first in the small garden at Horsley but more especially in the more ample grounds here in Albury. There must have been an influence, also, on the hundreds of visitors to our two garden menageries, especially the one at Albury. At least five other people who worked here as assistants for shorter or longer periods of time have gone on to take their places in the publishing world and to become authors in their own right. There is also the impalpable influence on many who have spent pleasant hours in this garden in whom, as a consequence, has been engendered an active sympathy for the preservation of our wild life.

It is not too much to say that our lives were occupied from the time we woke up in the morning to the time we went to bed at night almost exclusively with the animals we had gathered around us. In a way this created a vicious spiral. The more animals we had – and several thousands have passed through our hands – the more it cost us in money and to defray these costs meant more output in journalism and authorship, the fees from which met our costs. We never received a grant from anybody and we never asked for one. In the late 1940s I had developed a passion for journalism and for writing books. There was nothing deliberate about this. It all came about by accident but it was cumulative. For one thing it made me retire early from the Natural History Museum in order to devote myself to writing and from 1946 to 1964 I contributed a million words to the *Illustrated London News* in a weekly article on the *World of Science* page. In 1949, I was invited to contribute a Saturday Nature Note for the *Daily Telegraph* and these weekly contributions now total in excess of 360,000 words. I have also written over eighty books on natural history.

While the writings brought in an income that helped to defray the costs of our zoo, the mere presence of the zoo helped to provide materials on which the writings were based. Moreover, the fact of having to produce 1000 to 1500 words each week for the *Illustrated London News*, meant that I was always seeking for copy, for something to write about, and this made me much more observant, much more ready to go out and study animals either in our zoo or in the countryside, or to write about them. In this way I built up a vast store of information, in the form of notes, published material and the books acquired for reference purposes, as well as the knowledge I could carry in my head.

On the question of finance, it would be unfair not to acknowledge that, as time went on, the sales of Jane's photographs contributed increasingly to defraying the cost of feeding and housing our large animal family.

Another consequence of having our zoo around us was the unique opportunities for research that came our way. There has probably been more research on anting, that puzzling phenomenon of bird behaviour which I shall describe in several of the following chapters, in our zoo than in any other institution in the world. The equally puzzling phenomenon of hedgehog self-anointing was not discovered in our tame hedgehogs but it was widely publicized from our observations and for the first time took its place in the English literature, to pass from virtual obscurity to become a well-known phenomenon. Our zoo helped to solve the mystery of the Surrey puma, when the belief spread through southern England that a puma or cougar was rampaging through the countryside scaring livestock and people alike. At the least, our house and garden served as a clearing house for inquiries that materially helped the police in their efforts to unravel the mystery.

From 1934 until today I have had an active interest in the Loch Ness monster, probably the most widely publicized animal mystery of all time. For twenty-seven years I maintained a belief in the reality of a large unknown animal in that Scottish loch. It was largely the research carried out in the garden at Weston House, into the behaviour of aquatic animals, that caused me to change my mind. It is my hope that soon it may be possible to publish these and subsidiary results and so put this tantalizing problem in its true perspective. This is the only attempt of which I am aware to take the problem 'into the laboratory'.

The chapters that follow are an attempt to bring together some of our adventures into natural history in our 'zoo at home'.

2 *Aristocratic owls*

Years ago, in my more energetic days, I used to delight in taking walks through the countryside, alone at night – and especially on clear, frosty, moonlit nights. One of the main attractions for me was to hear the owls calling, and sometimes, when the moon was very bright, to see the shadowy form of an owl flying overhead. Imagine my excitement, then, when I received a message by a roundabout route that somebody had two tame tawny owls, and wanted to find a good home for them.

The young woman lived in a built-up area of Slough – an industrial town near London – but she had found the birds when they were owlets, somewhere out in the country, on the ground under a large tree. One had a deformed foot, the other was blind or partially blind in one eye, for the pupil was clouded over. She took them home and fed them and now, as nearly mature birds, she had had to swallow her sorrow and get rid of them because the neighbours were complaining of the noise they made. The circumstances suggested they were siblings that had been abandoned by their parents and, in passing, it is interesting to note how often one finds baby birds ejected from a nest that are deformed or sick. It leads to the suspicion that the parents are able to detect whether one of their offspring is a non-starter in life's race and deliberately cast it out.

I went to see the owls one evening. The young woman lived in one of a row of small terraced houses; the aviary was in a corner of a backyard smaller than the room in which I am sitting writing this, and that is eighteen feet by ten. A high brick wall ran along the backs of the backyards. With houses so close together, backed by this brick wall, the sound of an owl hooting would have reverberated back and forth, unable to escape. Two owls hooting, especially in the dead of night, must have been near to bedlam.

No wonder the neighbours complained.

At the time, I had a rooted aversion – in principle – to keeping animals in close captivity. However, the early history of the owls showed that to liberate them would have been to pass sentence of death on them. This and the fact that we could provide a roomy home for them and have the pleasure of at least making the close acquaintance of these birds, overcame any scruples I may have had. We built an aviary especially for them, with posts driven into the ground and rails across making a framework eight feet high by twenty feet long and ten feet wide. This we covered with wire-netting and inside we fixed branches decorated with renewable evergreen foliage, to give something of a natural habitat. The important thing was that the owls had sufficient room to be able to take short flights and really spread their wings. In the event both of them lived for a long time, one for nearly fifteen years.

A solitary robin was sitting on the roof of the aviary the evening the owls arrived, and he uttered his alarm note at the sight of them. The next morning it was a very different story. I was up at dawn and my first anxiety was how the owls were faring. I was in time to witness the beginning of a natural drama: a blackbird foraging on the ground caught sight of the owls and uttered his noisy alarm note and in a surprisingly short time the foliage of the tree sheltering the aviary was alive with small birds that had flown in from all directions. There must have been at least fifty birds in the branches of that tree – blackbirds, song-thrushes, chaffinches, sparrows, robins, wrens, blue and great tits – and the noise of their alarm chorus was unbelievably loud. As one who has always preferred to be at peace with his neighbours, I felt more than a tingle of apprehension about the outcome if we were to have this din repeated each morning. It seemed as if this dawn hymn of hate must awaken the whole neighbourhood.

'What are we going to do?' I said to Jane, who meantime had joined me to share in the fun. 'If it goes on like this each morning, we are going to have our neighbours complaining, too.'

'Let them complain,' said Jane, 'they can't do anything about it.' She was still no more than a slip of a girl but she already knew where she was going and what she wanted to do.

'Yes,' I countered, 'but supposing the owls' hooting keeps them awake at night and then, just as they are getting off to sleep, this hideous dawn chorus wakes them up again?'

Jane shrugged her shoulders. It was clear we were going to keep the birds. What neither of us knew then was how amiable our neighbours would prove to be, with one exception.

If the noise was quite extraordinary, the sight of the tree was even more so. It seemed literally alive, as if its foliage had assumed some phenomenal animation, for the birds moved continually from twig to twig in an excited General Post, calling vociferously in great agitation. Every now and then one would fly away over the hedges only to return a few minutes later. And for every one that flew away two or three more flew in. So it went on for three-quarters of an hour before the sounds finally died down.

The following morning there was the same excited assembly of small birds in our immediate neighbourhood and the same din from their alarm calls. Both began at almost precisely the same time by the clock and continued on this occasion for an hour. Matters had begun to seem more difficult than I had foreseen, but there was comfort in one thought: the mobbing of owls by small birds has been shown to be an automatic or innate reaction at the sight of an owl. Such reactions tend to grow weaker with repetition, so there was hope that in time the small birds around might become accustomed to the presence of the owls and cease to make this fuss. At all events, it would be worthwhile taking careful note of what happened.

Within a week there were signs of a diminution in the number of birds participating and, naturally, in the volume of sound they produced. The duration of activity shortened, the participants were somewhat less agitated, and they did not form such a tight group in the tree. Instead, they were more spread out over the garden. Their individual alarm notes were just as loud but they were less sustained, and the chorus itself was intermittent instead of being continuous as it had been in those first few days. If we can describe the chorus on that memorable first morning as a heavy black blot, we could say it had become no more than a smudge by the end of the week.

A few weeks later the morning visitation had virtually ceased. What is more, during the twenty years that followed it was never repeated with anything like the same force that it had on the first few days after the owls had come to live with us. It happened occasionally, at rare and infrequent intervals, that a blackbird coming near the aviary might sound its alarm notes, calling forth a few alarms from other birds around. But I have even seen a bird, feeding on the ground, make its own alarm call in response to such an alarm note without in any sense interrupting its feeding.

Someone who studied this matter came to the conclusion that small songbirds have the innate reaction to scream alarm at the shape of an owl. Putting this another way round, it means that when a songbird sees an owl, or anything in the shape of an owl, it cannot help drawing near to sound the alarm. However, if this were all there was to it we would have to explain why I have several times seen a blackbird sound its alarm call in response to one of our owls calling, although it was clear the blackbird could not see the owl. We would also have to explain why the stone owls on the gate pillars at Weston House, where we later went to live, do not alarm small birds.

The entrance to our present house is through a pair of heavy seventeenth-century iron gates hung on two stout brick pillars. Surmounting each pillar is a figure of an owl carved in stone. These are sufficiently lifelike that, were they coloured to represent an owl's plumage, they would be taken for live owls by everyone passing the gates. As it is I have never, in the twenty years we have been here, heard a single bird raise its alarm call or seen any sign of the stone owls being mobbed. On the contrary, many times I have seen birds perching unconcernedly on the tops of the brick pillars, even on the head of one or other of the owls. So shape cannot be everything. Probably the contrasting colour of the eyes is important and this would be lacking in a stone carving.

Once when my grandson, Mark, was looking into the tawny owls' aviary, one of the owls flew from its perch and landed, clinging to the wire-netting, immediately in front of the boy's face. Had the netting not been there it would have landed on his face, perhaps with dire consequences. Mark was wearing a brown anorak at the time, with the rounded hood drawn tightly over his head. This, with his large, rounded, boyish eyes, gave him a distinctly owlish look. Perhaps the owl was taken in by this and mistook him for a rival. On the other hand, small songbirds have never attempted to mob Mark!

It would be possible to spend a whole chapter discussing the possible reasons why birds mob owls and the anomalies and contradictions involved.

One fascinating feature is that the birds lose all sense of territory, enmity between species is forgotten and, like a crowd of people roused to frenzy by a loud-mouthed orator, they seem oblivious to everything else that is going on around them. The comparison is justified, for often the mobbing is stimulated by the raucous cries of a jay, the traditional enemy of small birds and one which, under other circumstances, they would flee from or combine to mob in its turn. In addition the bird-mobbers seem insensitive to danger. This has been known since the fifteenth century at least, when the practice was to lure songbirds to a hide by putting out a stuffed owl, or tethering a live one so that it could not fly away. The small birds were then netted or caught with birdlime on branches arranged around the decoy owl. The method is still used in parts of Europe to catch birds for the pot, although it is illegal in Britain; a modification of it is used in North America to catch birds for ringing.

The effect on the owl of having a screaming mob, threatening and jeering, in a quite bewildering manner is also interesting. To all intents and purposes the owl, the central figure in the drama, appears to be quite unmoved. It puts up with the noisy crowd for a while and then, seemingly unhurriedly, spreads its wings and flies away. Appearances are, however, deceptive.

Amid all the clamour and agitation the owl sits immobile, staring straight ahead without a movement of either of its three eyelids. When at last it does move, it launches itself into the air in exactly the same way and at precisely the same pace as it would if there were no disturbing element within miles of it. Yet, when one can observe the bird closely, it is possible to see that it is disturbed. One of the great advantages of having animals under restraint is that, over the years, they become used to one's presence and so tend to behave naturally, or at least without inhibitions. One can then get very close to watch them. On several occasions I have crept right up to the wire-netting of the aviary, where an owl was within two feet of my face, and then I could see that, in spite of its impassive appearance, there were small signs of unease. It began to look from one side to the other and move its feet uneasily on the perch. Having regard to the acuteness of an owl's hearing, by which it can home accurately on the rustle of a mouse in the grass, in pitch darkness, it could hardly be otherwise. Nevertheless, owls do not readily show signs of distress, as when a passing motor-car backfired when I was standing close to the owls, talking to them. Most birds would have reacted sharply to this loud, explosive sound, by flying up into the air. Our owls merely gave a sharp turn of the head in the direction of the sound, with an air of great concentration.

The noise made by the birds mobbing our owls was relatively short-lived, as we have seen, but that from the owls themselves was not! The tawny owl is liable to use its voice at any time of the year, and from January to June it is in full voice.

In the January following their installation at our house I returned home from London one evening just before midnight. As the train drew into the station and I started walking home I could hear an owl calling and two owls answering, more or less in unison. The sounds came from the direction of my house – and this was half-a-mile from the railway station. My heart sank. There was no doubt that what I was hearing was a wild owl in the woods opposite my house calling and receiving an instant reply from our two beloved orphans. At each step on my way down the village street, the owls' conversation grew more startlingly clear on the still night air, strident in contrast to the silence of the darkened houses on either side of me.

You can silence a day-flying bird simply by shutting it in the dark or throwing something dark over its cage. You can't 'throw something dark' over an aviary and, in any event, owls are at their most vociferous in darkness.

So I went to bed.

Two days later I received a letter. It was from a titled gentleman who had only recently come to live in a house a hundred yards from ours, on the other side of the road. It read:

> I understand that the owls which keep me awake at night live in your garden. I shall look forward to hearing that you are going to put an end to this nuisance . . .

My reply was as courteous as I could make it, but I tried to point out that my tame owls were making no more noise than the wild ones in the woods that backed onto his garden, and were therefore nearer to his house – and that possibly it was these that kept him awake more than mine did.

I knew my defence was somewhat thin and felt guilty about the whole affair. It was no surprise to me therefore to receive a further letter, two days later:

> Dear Sir,
> Now that we have established that you are responsible for the nuisance we can decide what further steps to take.

There followed a thinly-veiled threat of legal action, so I telephoned my solicitor, telling him of the incident.

'Take no notice,' he said. 'There is no law against keeping birds. Just do nothing.'

'But I must reply to this letter,' I pleaded.

'All right. Take this down.'

My solicitor dictated over the telephone a letter of two sentences. It was masterly. I did not keep a copy of it but I do recall that the words, while more or less incomprehensible to me, seemed to cover every contingency and to be unanswerable.

I duly wrote the letter. There never was a reply to it.

About a week later I happened to be up at dawn and I heard a chorus of owls from somewhere in the direction of the woods. My owls were silent.

I went outside to listen. In a circle of trees that surrounded the house of my titled neighbour were half a dozen owls, approximately one to each tree. Each owl called in turn, so their hooting went round and round in a circle, with my neighbour's house at the centre.

Our more immediate neighbours were aware that there had been this slight altercation by correspondence because I had made it my business to find out if my owls were being a nuisance. Without exception, all replied in the negative and said that not only were the birds not a nuisance to them, they actually enjoyed hearing them. It was impossible, in making these inquiries, not to disclose the identity of the one who had made the complaint, so when they heard the chorus of wild owls ringing his house, as I did, they were highly amused.

The whole affair was the joke of the village for some days.

Their general behaviour always makes me think of owls as the aristocrats of birds. This is not surprising since the wisdom of the owl has been proverbial for thousands of years; but the owl has, at the same time, the reputation of being a bird of ill-omen. Close and prolonged acquaintance with it certainly tends to strengthen the impression of wisdom. This may be the outcome of three things: its large eyes with their steady gaze; the absence of fussiness in its demeanour; and a very noticeable absence from its behaviour of what, for want of a better term, we call playfulness. This gives a calmness and an apparent dignity to all owls which is reminiscent of the human attribute of control of the emotions.

An owl can express its moods without the posturing, the spread wings, the fanned tail and the generally vigorous movements of most other birds. In short, an owl has poise. It can also put on a haughty look of the kind popularly associated with the upper crust of human society, especially in what is called the attenuated attitude. Our owls would adopt this attitude whenever a stranger, even a dog, passed the aviary. From the normal shape, with the disproportionately large head on the rounded body, the bird draws itself up to its full height so that body and head together become slender and cylindrical. At the same time the large facial discs of radiating feathers are contracted and the large eyes are narrowed to slanting slits, giving a mongoloid appearance to the face. The whole appearance is one of disapproval with a touch of the sinister. Our owls would adopt this attenuated attitude if someone with whom they were familiar went by carrying something unfamiliar or obnoxious to them. If I went by pushing a wheelbarrow, for example, I would see that the owls, which a moment before had appeared their normal rounded selves, were now shrunk into the attenuated attitude.

A family of my acquaintance that had a pair of owls always referred to it as the 'chinaman position'. This is something of a misnomer since 'chinaman' is an eighteenth-century name for a man that sells china, but it conveys the idea better than 'attenuated attitude'. It was also interesting that if one walked strangers past an aviary and the owls in it took up the chinaman position, the likelihood was that the visitors would fail to notice the birds. This says much for the camouflage value of the posture.

The shrinking is merely a matter of drawing in the feathers, for the actual body of an owl is ridiculously small and skinny. I once came across the carcase of a tawny owl in a meadow. The slight putrefaction of the flesh had loosened the feathers and recent strong winds had blown most of them away. The skinny body and scrawny neck looked ludicrously small compared with the apparent size of the live owl.

The attenuated attitude is only one of the many changes that can take place in an owl's appearance, although it is the most pronounced, and all these alterations take place rapidly and unobtrusively, with the speed of changing expressions on the human face. Indeed, one feels that owls are, in their expression of mood, all face. In none of the changes, however, is there any loss of dignity. It is the same in the matter of food.

Most hand-fed birds show an eagerness at the sight of food, expressed in a fair amount of bodily movement, and this is true whether they are hungry or not. A tawny owl has to be ravenous before it will show eagerness and even this is expressed by no more than leaning forward a trifle – although the final taking of the food, it is true, is a slightly undignified snatch with the beak. When the owl is just ordinarily hungry it will take proffered food with a magnificent condescension. If for any reason it is not ready to feed it will react to food placed in front of its beak by drawing itself up in what can only be described as a haughty manner. Should one persist in holding the food there in spite of this, the owl merely turns its head through 180 degrees, the owl equivalent of our turning our back on something unpleasant.

Perhaps we were more aware of these things because our tawny owls never learned to feed themselves. They had been hand-fed before they came to us and they kept this up, so that for nearly twenty years somebody had to go into the aviary twice a day and feed them with strips of raw meat – a total of nearly 14,000 feeding sessions – so it is no wonder we got to know these particular pets well. If food was put on a plate or in a bowl in the aviary, as was done with our other birds, they made no attempt to come down to it, and nothing we could do to persuade them otherwise had any effect. It is known that young owls leaving their parents are slow to learn to feed themselves, and that this is one of the main factors in the heavy mortality rate among young owls in the wild.

There were occasions, and these after dark, when one of the owls had been sitting on a branch with its eyes closed and had taken food without even lifting an eyelid. No doubt when an owl has to find its own food it shows a little more animation; but in our tawnies, habitually hand-fed, their actions underlined the fundamental immobility which is one of their main characteristics.

It must have contributed to these two owls' inability to feed themselves that we fed them with raw meat instead of their natural living food, which is mainly mice, voles, shrews and small birds. Once you dedicate your time and energy to rescuing and hospitalizing wild animals you quickly become squeamish about taking life, even the life of vermin. We never could bring ourselves to trap even the troublesome mice merely to feed owls. This had one benefit at least: we were able to prove wrong a deeply-entrenched notion, that owls need the fur and feathers on their food to keep healthy.

I remember one lady writing to me about her pet owl, which she had for some years. She had had to feed it on raw meat and, because she understood they needed the roughage of fur or feathers, she had conscientiously provided this, cutting up an old fur coat into small pieces and wrapping each piece of meat in a small fragment of fur. Like us, she had fed it any dead mouse she had found, but cutting up the fur coat had indeed been a real labour of love.

We were brought dead mice by our neighbours from time to time so we did at least have the occasional experience of witnessing an owl actually regurgitating a pellet of fur and bones. The first time one sees this it is somewhat distressing. The bird appears to be afflicted with paroxysms that shake its frame to its foundations. Then it opens wide its beak, leans forward, gives a final convulsive heave and retches a slimy mass about two inches long. But our owls remained healthy and lived long without having to suffer this daily.

I learned another thing from my years of hand-feeding the owls. There were times when one of them refused absolutely to take the food, even though it was being offered at the usual feeding time. It was useless trying to persuade it to accept the meat by putting it against the beak: the owl would turn its head away and if one persisted it would spread its wings and fly off to another perch. But I found that if one touched one of its talons with meat it immediately bent down, took the meat in its beak and gulped it. This makes sense for, when feeding, an owl seizes its prey with its talons and immediately transfers it to its beak. So there is an automatic sequence: touch with talons, lower the beak to seize.

Although essentially nocturnal the tawny owl is not above sunbathing, so in siting the aviary under a tree, to give the shade most owls seek by day, we made sure that at least one corner would receive sun. In sunbathing there is again a strong contrast between owls and the general run of birds. In owls it is carried out without any fuss or any of the exaggerated postures typical of most birds. The owl sits half-on to the sun with the eye that is exposed to the full glare of its rays shut, the other open. This position will be held for long periods of time in a pose as statuesque and static as though the bird was dead and stuffed.

In view of what has been said about the owl's dignified ways, we should hardly expect to find an unbending to the point of being playful. The score or so of other birds in the various aviaries around the owl's aviary were habitually busy all day long, except for a siesta on sunny days. According to the species and their differing dispositions, they were turning over the earth, searching among dead leaves, picking up pieces of stick or leaf, carrying one piece here, another piece there, making a pass at a neighbouring bird or displaying in a more courtly manner. At all events, they were busy all the time doing something, and doing something with some thing. That is, there was a general tendency to search, to pull to pieces, to pick up and transport, to accumulate materials in one spot, to be attracted by bright or coloured objects. If nothing else, they were forever preening, in short bouts or long sessions.

This is, of course, a general statement, the whole of which is not applicable to any one species of bird. It represents a synthesis of their appetitive behaviour, as it is called. An extension to it at any one point, or at several points combined, can lead to more familiar actions. Thus, pulling materials to pieces, whether leaves, bark or grass, and their accumulation at one site, can form the basis of nest-building. The attraction of bright or coloured objects could be the basis of the search for food, such as seeds or berries. The storing of these same objects would lead to food hoarding.

We saw little of this appetitive behaviour in our owls, by day or by night. It was as though their whole existence was divided almost entirely between taking their food and resting in the dignified posture of a born aristocrat. There was a slight departure from this occasionally, however. In one corner of the aviary we had fixed a wooden box high up from the ground, with one side open. It was the best we could do to simulate a hollow tree, in which the owls could shelter. A piece of sacking was thrown over it to keep out the rain. From time to time we would see one of the owls pulling off pieces of hessian. Occasionally, also, we would see them picking leaves off the conifer branches we fixed in the aviary. But even these departures from their normal immobility were done with slow, dignified deliberation. It would appear to be significant that tawny owls, as with others of their kind, use no nesting materials, or do so exceptionally. They may occasionally use an old nest of some other species, such as a crow, rook, magpie or squirrel, but that is no more than a substitute for a natural cavity in a tree trunk.

Soon after the owls were installed in their aviary I began to notice that whenever I walked by they would indulge in bill-snapping. That is, their mandibles would move in a regular way, making a sound similar to the one we make when something goes mildly wrong, a sound that is usually represented in writing by *tch-tch*. We make the sound by putting the tip of the tongue behind the upper front teeth and then moving it up and down rhythmically.

I soon took to responding to the bill-snapping by tch-tching. From then on I would always greet them with this sound, and if they were not making their bill-snapping noise as I approached they never failed to respond to my *tch-tch* with this soft, euphonious and, I am confident, friendly greeting. There was, in fact, only one occasion when it did not happen. Instead, one of them made the same kind of bill-snapping but at a slower rate, giving a more deliberate sound, at about half the speed. The surprising thing was that by reducing the speed this friendly greeting was so obviously converted to a menacing sound, just as we, by speaking slowly and deliberately, convey menace in the voice.

Menacing and sinister it was, for immediately it started this slow bill-snapping the owl flew at my head and left me with two pin points of blood, one on the upper lid of my left eye, the other on the lower lid, where its talons had penetrated. I have thought about it often and have never been able to account for this sudden attack which, as I realized the moment it had happened, came very near to costing me the sight of my left eye. It holds a general warning for anyone dealing with animals, that no matter how friendly they may be, and for however long they may have been friendly, there is always the risk of unpredictable attack. One cannot be too careful.

I have spoken of these sounds as bill-snapping, which is the usual name for it. It sounds as if the owl is opening and shutting the beak, snapping the two halves of the beak, the mandibles, together to produce a mechanical sound. Moreover, the mandibles move precisely in time with the sound. There is some excuse, therefore, for assuming the sound is mechanical, that is, a snapping together of the mandibles, and not vocal, and so it is of interest that H. Stadler, who made the most complete study of the vocalizations of European owls, maintained that the bill-snapping is vocal. In volume 2 of Witherby's *Handbook of British Birds*, page 339, his work is referred to, but an editor's footnote remarks 'we cannot accept this view' that bill-snapping is vocal.

I regard this conflict between Stadler and orthodox ornithological opinion as containing a fundamental principle, a warning that appearances can be deceptive in quite surprising ways. In this instance, any reasonable person watching an owl bill-snapping would, without hesitation, say: 'Quite obviously the bird is making the noise with its beak. I can see the beak opening and shutting in time with the sound and the sound itself is what one would expect if the mandibles were hitting each other.' This same reasonable person would say, with the editor of the *Handbook of British Birds*, that he would not accept Stadler's assertion.

However, I noticed early on in our experience with the two owls that the sound of supposed bill-snapping would continue after the bird had accepted a piece of meat and it was still lying crosswise in its beak. So there was no possibility of the mandibles touching each other, nor was there any movement in the mandibles themselves. It was the same the next time I gave it a dead mouse. I mentioned this to Jane and she agreed that she had noticed the same thing; and after this we both paid particular attention – and confirmed our suspicions.

The elucidation of the nature of the bill-snapping received more dramatic confirmation when we later visited a friend who kept several species of owls in aviaries. One was the large eagle owl and it had lost the whole of its beak and could only survive by being hand-fed. Yet it made the unmistakable and quite characteristic sound of bill-snapping!

Perhaps another correction to a deeply-rooted idea may be permitted. It is always said that an owl flies silently and that this enables it to approach its prey unheard. On many occasions I have stood in the aviary as the owl has flown past me, and I have heard a faint swish of its wings. Mice have more acute hearing than I have, especially for faint sounds, so I am quite sure a mouse would hear this noise and from a greater distance.

Another small discovery about the owl, made by Jane, is of interest. Up until then – 1956 – it appeared to have escaped the notice of ornithologists that an owl's large eyes are probably protected by a device almost invisible to the human naked eye.

When I put the owls into their aviary, on first arrival, one of them flew straight up, but instead of hitting the wire-netting ceiling it skilfully turned when practically at the wire and hung head down by its toes, almost like a bat. After seeing this I often pondered by what sense the owl made its split-second appreciation of the wire. It may have been by sight alone – it was half-light in the evening – or it might have been by some other sense.

It was nearly a year later that Jane drew my attention to halos of fine hair-like feathers surrounding an owl's face. Although the hairs are very numerous and project a full inch from the face, it was only possible to see them when the sun's rays strike them at a particular angle. Even then they are only just visible. It seemed likely that the halos might function like a cat's whiskers in the dark.

While pondering this problem I happened to be moving through some scrub when I brushed past a young birch. A catkin swung and hit me on the centre of the eyeball, in the middle of the pupil. For a few moments the pain was excruciating. As a rule we are protected by the reflex blinking of the eyelid but here was a rare occasion when the speed of the object was faster than the blinking of the eyelid. It made me wonder how animals fare in this matter, especially owls, with their proportionately large eyes, for they habitually fly through the foliage of trees.

Much has been made in this chapter of the dignified bearing of owls, which must have contributed to the notion of their 'wisdom'. But as I mentioned earlier, owls have also earned the reputation of being birds of ill-omen. This is hard to understand until we study carefully the attenuated attitude in which the eyes are closed to slanting slits reminiscent of the eyes of Mephistopheles. This is even more pronounced in short-eared and long-eared owls disturbed at the nest. They do not then display the attenuated posture but spread their wings, as if covering their brood – and the face looks positively satanic. Moreover, in addition to the slanting slits of the eyes there are the two 'ears', tufts of feathers on top of the head where the devil's horns are supposed to be.

It is said that the attenuated attitude gives an owl increased camouflage, especially because it makes the eyes inconspicuous, but it is probably of greater interest that it helps us to answer the question of how far owls can see in daylight. Many times I have walked up the path to my owls' aviary to see them in the attenuated attitude as I came into view round a shrubbery. Once they recognized me their feathers quickly fluffed out, the satanic slits became rounded eyes and the owls were back to normal.

I made numerous tests, with a variety of noises, while watching the owls through the shrubs. These included a particularly noisy alarm clock. The owls turned their heads sharply in the direction of each sound, but it was only when I stood up suddenly to reveal myself that the owls became attenuated. Always they fluffed out again as soon as they recognized me. These and other tests left me in no doubt that attenuation in owls is a response to some unfamiliar sight – hearing evidently has nothing to do with it and gives me the rough guide that my tawny owls failed to recognize even large familiar objects at distances over thirty feet.

In due course, both owls went the way of all flesh. One died at the age of ten. It was found one morning on the floor of the aviary, apparently dead from natural causes. The second survived for another four years and was found in almost identical circumstances. But memory lingers on. A year or two later the aviary was dismantled, its woodwork rotted, its wire-netting rusted. The floor of the aviary was dug over and planted. But the resultant flower-bed is still referred to as the owl's aviary.

3 *The lass with the delicate ears*

Shortly after Christmas 1954, a somewhat unusual letter arrived.

'Dear Dr Burton,' it read. 'Would you like to adopt a young genet?'

I was recuperating from 'flu at the time, and my resistance must have been weakened.

'Dear Miss Lindley,' I replied, 'I would very much like to consider accepting your genet . . .'

Genets belong to the same family as mongooses, although they look more like tabby cats. They range throughout Africa and, although they have been introduced into the south of France, this was our first prospect of looking after an 'exotic' animal. It had been known by several names before it came to us but we promptly named it Jennie and, as it turned out, she was a female.

Jennie had been rescued, when a suckling, with an injured spine. An African had given her to Marie Lindley, then practising as a surveyor in Uganda, and she had nursed it, hand-fed it to first maturity, and then brought it to England on her next leave. When she found it was not practicable to take it back she wrote to me.

Since Miss Lindley had always intended to release the genet into the bush, once it had fully recovered from its injury, she had made no attempt to tame it fully. I clearly recall the first time I set eyes on the genet. Marie Lindley brought it in its travelling box to my room in the Natural History Museum. When she raised the lid the genet lifted its head to look over the edge of the box. 'What a beautiful animal,' I exclaimed, only to be rewarded with a show of business-like teeth and a throaty hiss. Marie could handle the genet but it was clear from this first meeting that nobody else would be able to do so with impunity.

In fact, Jennie never became more than semi-tame with us. She would climb onto our shoulders, sniff our ears, even settle down for a while on the back of my neck when I was at my desk writing; but any move to pick her up was quickly met with the display of teeth and the hiss.

In anticipation of Jennie's arrival I had fitted out a small room next to my study with light boughs to give some resemblance to a woodland, the natural habitat of genets. Also, since this one came from tropical Africa, a greenhouse heater was installed. We included in addition a tray of sawdust, for a genet is as clean as a cat.

Jennie was indeed beautiful – about eighteen inches in total length, of which nearly half was tail, her slender body was on short delicate legs, and her head was patterned with lines of brownish-black blotches on a silver-grey background. There was a conspicuous white triangle under each eye and a black line along the middle of the back. Her tail was ringed in black and white.

When I arrived home I took the box into the room we had prepared, unfastened the hinged lid and made to raise it. At that moment it was thrust upwards from within and a streak of mottled lightning leapt upwards. The nearest vertical object was the window and the genet was carried up to this by the force of her leap. For an infinitesimal moment she clung to the glass with her claws before, with a twist of the body, she changed direction and landed on a horizontal bough. She stayed there a few moments, then she started to explore her new world.

Silently, almost snake-like, she walked the length of one bough after another. It was a slow progression. She would put forward first one forepaw, then the other. When sure that they held firm she lifted first one hindpaw then the other. Slowly, slowly, she made her way. All the time her neck was stretched to the full and her head moved from side to side, her ears were twitching, her eyes fully concentrated and her nostrils working. Every detail of her new world was inspected minutely, even to the occasional cobwebs and blemishes on the walls and ceiling. Nothing seemed to escape her scrutiny. It took her a long time to complete the circuit of the room and its furnishing of boughs, and I stood watching, entranced.

Having completed the circuit she went over the whole again, following exactly the same route, but this time slightly more quickly, although she was still testing with her feet and inspecting everything with nose, eyes and ears. Finally she went round a third time – but this third traverse was carried out at speed.

I formed the impression that in the first and second journeys she had been registering and memorizing every detail of her new environment. This seemed to be borne out by her subsequent behaviour, for always thereafter she ran and jumped rapidly all round the room. Her room was immediately under my bedroom and at night I could hear her, as I lay in bed, racing round the room at top speed.

It seems a reasonable assumption that a nocturnal animal, like the genet, must carry a memory of its environment so detailed that it can move about in pitch blackness yet know exactly where it is going. Confirmatory evidence came later when we built a wire-netting enclosure – in effect a large aviary – outside this room, in the garden. When this was ready I opened the window and Jennie went out. She found herself in a 'room' bounded on all sides and on the ceiling by wire-netting, with a system of boughs reaching to all points in the room.

Within these outdoor quarters Jennie first went round at the same speed as she had made her first inspection of the room. Again, she inspected everything with eyes, nose and ears. Having completed the first circuit, she followed the same path at a slightly increased pace, and then, on her third tour raced around it at top speed. When we moved to Albury from Horsley, where we first had the genet, she treated her new quarters in exactly the same way. Here, also, my bedroom was directly over the genet's quarters and for the next seventeen years I could hear her at night racing around her cage at top speed, night after night, in what must often have been total darkness. She became less sprightly as she neared her last years, but she never seemed to falter and only once to my knowledge missed her footing.

It was when she was making the first circuit of her outdoor quarters at Horsley that this occurred. She was walking slowly along a branch when, in spite of the care she was taking, she missed her footing, slipped and hung from the branch by all four paws, just for a moment, before swinging herself once more up onto the branch; she could be described as having, inadvertently, performed a somersault at this point. Always thereafter, whenever she reached this same point, she would somersault instead of running straight along the branch. I took this to mean that so accurate was her memory of her surroundings, from these first inspections, that even an error on her part had to be faithfully repeated to keep the sequence intact.

During her first inspections of her original quarters at Horsley, while we were still total strangers to her, Jennie initiated us into the details of a genet's aggressive displays. (I use the plural because Jane was also present, as she usually was when I was making special observations: we worked as a team, corroborating or correcting each other's observations.) If either of us, Jane or myself, moved suddenly or went too near Jennie while she was inspecting the room, she favoured us with her aggressive display. She would arch her back, hold her head low and the line of black hair along the back would be erected in a crest. At the same time the tail, which was always held straight out behind her when she was walking or running, and in repose was quite slender, became a fair representation of a bottle-brush.

Another ingredient of the aggressive display was vocal. From somewhere inside the genet came a simplified version of a cat's purring, perhaps better described as the sound of an old-time kettle boiling. This seemed to be used rather as a warning that her patience was nearly exhausted, that one more provocative act and she would spring into the attack. But it was the crest that held the greatest significance.

When in full display a genet's crest extends from just behind the head and along the back to the root of the tail. It is important to say, however, that the full display is not necessarily given for each alarm. At a slight disturbance only a few hairs would rise, at a point midway along the back. As the disturbance, and the genet's consequent alarm, increased more hairs, in front of and behind this small bunch, became erect until finally the crest was continuous. This gave us a very useful yardstick for measuring the degree of alarm experienced by the genet at a given moment.

One day, soon after her arrival, Jennie seemed to be particularly nervous. When I went into her room she was on a branch just above the level of my head. Instinctively, as we say, I put up my hand to her in a misguided attempt to soothe her. She went into her full aggressive display, crest, bottle-brush and all. In addition, she bared her teeth and treated me to a low hiss. It was, although I was not really aware of this at the time, the hand that was at fault.

To an animal the human hand has a particular significance. It is a source of danger to the untamed animal, whereas in one fully tamed it is a source of security, food or petting. It seemed to me that I had an opportunity to test this for myself with the genet. A few days later, when she seemed to be in a relaxed mood, I carried out simple tests, using the index afforded by the movements in her crest.

As might be expected, if I thrust my hand quickly towards her she gave me the full aggressive display and had my hand come within reach of her teeth she would undoubtedly have bitten it.

Even when I moved my hand towards her slowly, her crest came up, beginning at the middle and extending in each direction to the tail and the head, in proportion to the nearness of the hand to her. From further observations it was clear that these two sequences were invariable, and that movements of one's hands, even when not deliberately directed at her, always evoked a response.

I carried the matter further to see if it was merely bodily movements of any kind or only movements of the hands that evoked an adverse response. Keeping my hands in my pockets, I tried moving my head towards her. There was little or no movement in the crest, even when I jerked my head violently towards her or thrust it as close to her as my hand had been previously. So long as my hands were out of sight, in my pockets, I could move my head or body towards her or bring my knee up to her and she showed no alarm.

It could be argued that merely because the genet had been handled, however benevolently, she viewed hands with disfavour if only because they represented loss of liberty. All I had been doing was confirming the findings of other people with four-footed beasts. There was, for example, the occasion when a group of scientists engaged on agricultural research visited a farm. A dog was chained near the entrance to the yard and the farmer warned the visitors that it was 'so savage we can do nothing with it'. The scientists went past the dog, into the farmyard, and went into a huddle, discussing what they were going to do. After a while one of them looked up. He saw one of their party had gone back towards the dog, with his hands in his pockets, and was speaking soothingly to it. From being a barking, snarling dog with hackles up, they saw the dog slowly drop all signs of aggression. It began to wag its tail and, in the end, the scientist, having quieted the dog and won its confidence, was seen to take his hands out of his pockets and fondle this savage, unmanageable dog, who responded with the usual canine tricks of affection.

The contrast between the savage farm dog and our genet underlines the difference between a domesticated species and one that is truly wild. Graceful as a ballerina, agile, supple and muscular, it was as though all Jennie's nervous tissues and energies were used solely to serve her physical needs with none left for intelligence. At any rate, she never learned to discriminate between the hand raised in offence – not that we ever raised a hand to her offensively – and one bringing her food. In short, she was as beautiful as a dream and almost as senseless.

Although we took care never deliberately to provoke her, there were times when we were compelled to put a hand near her, as for example when putting her food bowl down in its appointed place. Often, when she was more hungry than usual, she would run towards this spot, when one went in with her food bowl, and station herself there ready to feed. Then, as your hand went towards her – a necessary manoeuvre since she was there first – she would bare her teeth, hiss, show the bottle-brush and imitate a kettle on the boil. On such occasions, even when one's hand had been withdrawn leaving the bowl in position, she would continue her kettle-boiling while eating.

Of course, it had to happen sooner or later. There was the day when I took her food in as usual and as the bowl touched the floor she bit me, in the fleshy part of the thumb. Blood immediately poured out from that vulnerable area but a couple of minutes with the hand held under the cold tap and the flow was staunched. The only sign of my injury was a barely visible puncture in the skin from a needle-like canine.

It would be wrong to represent the genet as wholly reprehensible. One had only to stand for a while within her domain, keeping the hands still, of course, and she would soon jump or climb onto one's shoulders. Then she would rub her flanks against the back of your head and gently inspect your ears and lips with her delicate nose. The line between this endearing show of affection and her display of aggression was a very thin one. Raise your hand suddenly and the likelihood was that she would stretch out her long neck and strike in a snake-like manner with her teeth, perhaps only butting your hand with her front teeth in warning.

Zoologists studying wild animals in the field are coming to the conclusion that when they attack human beings it is primarily because they regard them as intruders into their territory. Unless I read more into the incident than is justifiable, we had a prime example of just how strong is the attachment to territory.

There came a time when we needed space for another purpose and there was no solution to our problem other than to take part of Jennie's room. It was unfortunate, too, that this necessitated using that part of her room containing the cupboard on the top of which her nesting-box was situated. Not more than a third of her room was needed, however, and this left what one would have thought was ample space for her. A wooden framework with a door was fitted across the room and over this was stretched the usual wire-netting. Her view was, therefore, unobstructed, so that even though she could not wander freely where she had previously been able to, she could still gaze upon what had formerly been the rest of her territory. We moved everything with which she had been familiar into the remainder of her living space and her sleeping-box was sited at the same height and in a similar position.

It would probably have been better, had it been possible, to move the genet to entirely new quarters. She would have been upset at first, no doubt, but in time she would have settled down, making every corner of her new quarters her own. She would, we may suppose, have forgotten in due course all about her previous home. As things stood, she appeared not to forgive us for dispossessing her of part of her territory and to remain seemingly resentful.

To speak of an animal withholding forgiveness, or showing resentment, may shock the purists, but that is what it looked like, although it is not easy to give convincing or detailed evidence to prove it. As to showing resentment, we have at the moment a half-siamese cat in the household that tends to curl up in my favourite armchair. For the most part, I leave her to sleep there and choose another chair if I want to relax. There are times when circumstances demand that I should pick her up and put her on the floor, to use the chair myself. At such times she gives me a baleful look of such intensity that if it is not expressing resentment then it looks mighty like it.

So with Jennie, I am tempted to say that she looked resentful. But this may be subjective and I look for more solid arguments. The first is that from the time the partition was put up Jennie became markedly more 'touchy'. Whereas previously one could go into her room and be treated either with complete indifference or with some show of interest, if not affection, now merely to step into her territory was enough to call forth her hissing and kettle-boiling.

Another symptom of this supposed resentment lay in the habit she developed of stationing herself at the foot of the partition and spending long periods of time looking into the territory she had lost. Here, also, the evidence of the eye cannot easily be put into words except to say that as she gazed through the wire of the partition she wore an air of dejection, almost of melancholy and longing. Then, if one came in through the door in the partition she would hiss, boil or strike with her teeth at one's shoe but without leaving tooth-marks, to a degree much more marked than

we had noticed before. It was as if she identified her loss of territory with those of us who had been responsible for it; and more
especially did she show this displeasure when she happened to be
in the area bordering the partition which now cut her off from her
former haunts.

It is difficult enough to analyse with certainty the emotions and
intentions of our fellow-men. We may suspect that somebody
refuses to forgive and still resents some action on our part. It is
another matter to be sure of it. If we question that person about
it, his answers may not represent a statement of absolute truth;
and even he may not be able to say, if he is really honest, whether
or not he still resents and refuses to forgive. So, in our estimate of
such a situation as between two human beings, we have to be
content with the overall impressions we gain from what we see. In
the situation I have described, involving a non-human, the
chances of being more precise and of reaching absolute truth are
even more slender. After watching this situation over a period of
several months, I can only say that the genet reminded me of how
people behave if dispossessed of something they can still see, or of
how a nation behaves when it has lost a strip of territory to a
neighbour.

Even if it is not permissible to speak of a genet in terms of
showing forgiveness or resentment, the incident at least emphas-
izes the strength of the territorial instinct. An animal may have
taken over more space than is necessary for living, yet it will fight
to retain the whole. Deprived of even a corner of this, the impulse
remains to take the first opportunity to recover it. Whatever the
processes that foster this impulse are, they must, in an intensified
form, come remarkably near to what we mean by resentment.

The relationship between Jennie and ourselves was probably not
unique but it was certainly unusual. It may be that, because we
were unable to handle her, while at the same time enjoying the
bond of tolerance that existed between us, this caused us to use
our eyes more in studying than might have been the case had we
been able to pet her. This may also have been why I came to
notice her ears so closely.

Early in our acquaintance I nicknamed Jennie 'The lass with
the delicate ears', an unforgivable parody, perhaps, of 'The lass
with the delicate air'; and why not? She was a thing of beauty,
admired by all who saw her, and marvellously photogenic. For me,
her ears held an especial fascination: they were one of the first
features to catch my attention when she took that original tour of

inspection around her new home. And they interested me for two other reasons. In the first place they were large and had an intrinsic beauty, for each was like one of those delicate seashells that give one such aesthetic pleasure, and to which no photograph can do justice. But it also happened that, at that time, I particularly wanted to study animals' ears at close quarters.

For some time past, in popular lectures, I had expressed the view that it is possible to make a gross assessment of the habits of an animal merely by examining the visible structure of the sense-organs. Although I had put forward this view with seeming confidence so often, there had always been the lurking doubt in my mind that I may have been misleading my listeners. I had felt this especially in regard to the ears, because these organs so often, in addition to serving as organs of hearing, help to cool the body. So it may sometimes happen that large ears may be not so much a sign of acute hearing as of providing a large surface for the radiation of body heat. The large ears of the African elephant are a case in point: it lives more in the open than the Indian elephant, with its much smaller ears, and waves them slowly in the heat of the day to disperse the body-heat radiating from them.

Jennie's ears were large in proportion to her head, and although her whole face was alert, the muzzle fox-like and doubtless housing a keen sense of smell, there could be little doubt, from what I could see, that hearing was the most important of the genet's senses. Although large, Jennie's ears were delicate in texture and colour, semi-translucent and extremely mobile, giving an immediate impression of being highly sensitized. They were always on the move, turning simultaneously to the front or diverging to the sides, so that the gap between their inner margins, as measured across the forehead, was constantly altering. At such times, one had the impression that the ears were focusing or, perhaps more strictly, using a triangulation for judging direction and distance. Another feature was the way both of them would be swung round to the rear, presumably to detect and locate sounds from behind without turning the head.

In addition to their simultaneous use as a pair, the ears were often used independently, the one swinging to the side or the rear while the other was facing the front. So, as one watched the animal's head, the ears were seen to be constantly on the move, swinging this way and that, curving down or stretching fully upwards, as if to give their possessor a continuous record, a sound-picture so to speak, of its immediate environment.

Perhaps an even more remarkable feature of the genet's use of its ears was that they were continuously quivering, not in an even or regular manner such as we might associate with shivering due to cold or nervous excitement, but an irregular quivering which seemed to coincide with the variety of sounds reaching them. As a result no doubt of the habit I developed of tch-tching to the tawny owls, I tended to make the same sound to all the inhabitants of our small menagerie. It was when I faced Jennie at close quarters and made this noise that I noticed how her ears vibrated in perfect time with the sounds I was making. Moreover, the amplitude of the movements of her ear-flaps rose and fell as I made the sounds louder or softer.

There were therefore two components in the use of the ears: the gross movements for searching and locating and the minute movements, the quiverings, which were continuous, even when complete silence reigned, and which were synchronized with sounds when these occurred. Then came a further interesting point. Although the gross movements ceased when Jennie was asleep, the quivering did not. Although fully active from about 5 p.m. until the following morning and sleeping during what is for us the active part of the day, I never caught her asleep. She slept curled up like a cat; but no matter how gently I approached her sleeping-box, her eyes would be open and her head raised in an attitude of watchfulness. On the other hand, once she had identified me she would lower her head, her eyelids would slowly drop and, after having raised them again two or three times, as if to reassure herself, she would settle down to sleep once more, her eyes fully closed. She was fast asleep so far as I could judge yet her ears continued to quiver, reminiscent of the quivering of an insect's antennae. Putting together the fact that Jennie could not be caught napping and that when she did, so far as we could see, go fully to sleep the ears continued to quiver, it is reasonable to assume that the quivering continued throughout sleep and that the ears did not fully rest whatever the remaining senses might do.

It is well-nigh impossible for us, who rely so very much on the use of the eyes, to appreciate what so complete a use of the sense of hearing means. And being unversed in the science of acoustics I found it impossible to understand precisely the function of the quivering action in the genet's ears. On the other hand, I was fully satisfied that, in the case of the genet, large and mobile ears denote an acute, very delicate and highly efficient sense of hearing even if, at the same time, the ears do perform the additional task of providing a large surface for the radiation of body-heat, and acting as a sort of thermostat.

There was one particular episode that shed interesting light on Jennie's hearing. One day a fledgling bird was brought to me, said to have been abandoned by its parents. (Too often people pick up newly-fledged birds under the impression that they are abandoned, whereas if they were less precipitate and more observant they would discover the parents watching anxiously from cover for the interfering human to depart.) At all events, I decided to feed this one and it so happened I was standing near Jennie's home holding the fledgling in my hand. It gave one squawk and opened its beak to be fed. It was near midday and Jennie was fast asleep in her box eight feet up from the ground. The moment the fledgling squawked, and before the sounds of the squawk had died away, Jennie had awakened, leapt from the box, run down a branch to the floor of her room and across to the wire-netting at a point as nearly as she could get to where the fledgling was. At no time could she have seen the baby bird. It needed no more than the sound of its begging call to spur the genet into this lightning-like dash, almost too fast for the eye to follow. I needed no more to tell me that the main food of wild genets must be small birds. Certainly her agility in climbing and scrambling among the branches we put into the three different 'homes' she had during her stay with us suggested that nesting birds would be at risk where genets live.

I was also interested to see how Jennie, when in her outdoor enclosure, picked up the sound of a motor-cycle coming along the road past our house. She followed it with both ears cocked, moving her head in perfect timing with the vehicle, which she could not see. As the motor-cycle drew opposite us the rider changed gears with an appalling explosive sound. Immediately, her ears were turned to the rear, presumably cutting out the excessive noise.

It must surely have been by only a hair's breadth that genets missed the place now occupied by the domestic cat. In the south of France they are said to be tamed and kept as pets in houses to rid them of rats and mice. I used the phrase 'they are said' because I have never seen one in a house in France, and I am becoming very shy of accepting as true anything I have not seen or investigated at first-hand. My source of this information is in several books on animals, and these also tell me that the famous 'cats of Constantinople' were genets, and that genets were popular pets in past ages in Europe, where they were sold in the markets as early as the seventh century AD.

Genets, in my experience, are superlatively silent. They are said to miaow like a cat and to purr, but our tame genet never did more than hiss like a snake when anything annoyed it, and followed this with a sound that seemed to come from somewhere among its internal organs, a sort of liquid bubbling which I have already likened, appropriately, to that almost extinct sound of a kettle boiling on the hob.

Perhaps one of the main reasons why the genet ceased to be a popular and widespread household pet, and failed to occupy or usurp the titled, almost regal, position now held by *Felis catus*, is that it not only feeds on small rodents but has a penchant for poultry. That is something I can well believe, because one way in which we helped to feed our genet was by retrieving the carcases of birds we found lying on the road, casualties from collisions with cars. This may sound heartless, but at least it removed these pitiful relics from public gaze, and it enables me to say with certainty that our genet became more excited in the presence of a bundle of feathers than when confronted with a dead mouse or vole, brought to us by a neighbour whose well-fed cat had made a kill and brought it indoors to drop on the carpet.

The books also say that genets eat reptiles and insects. The first of these we never tried, but insects are clearly a delicacy. Our genet devoured the troublesome cockchafers with as much delight, and certainly more noise, than a gourmet does caviare. It also ate some kinds of fruit (grapes especially), needed a liberal ration of grass, and showed considerable eagerness for the taste of sugar. Raw egg was taken avidly, which may be another indication of habits that militated against a more widespread adoption of the genet as an animal about the house or the farmyard. In Majorca, where the genet has been introduced and gone feral, it is trapped with box traps, primarily as a persistent predator of poultry.

Jennie died in her sleep at the age of 21½ years, the previous record for her species being 12½ years. During her last two years of life she did little more than sleep and come out at regular times for her food. Even before that she was showing signs of old age but up to the last she still retained the ability to climb, when she so chose, although her old nimbleness had departed.

4 *Foxes' idyll*

In 1954 I was making an extensive search of the literature on foxes, mainly to obtain circumstantial evidence of their intelligence. I have never met a more tantalizing or exasperating piece of research. One writer after another made the usual enthusiastic remarks about what a magnificent animal the fox was; gave it the usual praise for its shrewdness, cunning, cleverness, intelligence or whatever word or words the particular author favoured, told the usual anecdotes of what a particular fox did under the stress of seeking escape or sanctuary from a pack of hounds – but none said anything more positive about the animal itself. Similarly, there were appreciative remarks about the attractive appearance and manners of the cubs, and descriptions of their playfulness, but again, little of solid worth.

As so often happens, fate took a hand. Early in the following year I saw an advertisement in a newspaper, offering a tame fox cub for sale. I wrote immediately, for here seemed the opportunity to study the animal for myself. It was one of the very few occasions that we bought a member of a wild species.

The cub arrived sooner than I had expected, and with it came a note saying it was not yet weaned. So I gave it some milk. The suddenness of its arrival also presented the urgent problem of where to house it. One of Jane's pet possessions at that time was a colourful gypsy wagon. Since she was away visiting I cleared the wagon of its few contents, covered its floor with peat litter, put in a sleeping-box, and another housing problem was solved, at least temporarily.

Then another difficulty arose. Although the cub had been fed it appeared restless, more restless than I would have supposed it would be merely through being in unfamiliar surroundings. An hour later it was still uneasy and still continued to call with a bird-like trill which every now and then ended in a piteous puppy-like whine. I sat on the bench in the gypsy wagon and looked down helplessly as the cub wandered around and over my feet, trilling and whining. I tried to imagine what a vixen would do in the circumstances, how she would interpret these signs. It was the cub that gave the answer. It took the cloth of my trouser leg in its mouth and tugged. It tried to scramble up onto my lap and when, finally, I had the wit to pick it up, it tried to chew the lapel of my jacket. It then dawned on me that the vixen, under such circumstances, would realize that the period of weaning was at hand and would go to fetch food. As I went towards the house for some meat the puppy-like whining turned to a small but unmistakable bark.

Once solid food had been brought there was no longer any doubt. The cub ate ravenously and, satiated, indicated once more that it wanted to be picked up. Nestled in my hands, it dropped its eyelids and soon allowed itself to be put in the sleeping-box, where we heard no more of it for several hours. That brief experience had taught me more than all my previous search through the literature.

There was an air of alertness about the cub, when it was awake, which was enhanced by its pricked ears. There was, too, a seeming purposiveness about its every movement, more pronounced than I could recall in any puppy or kitten. It appeared to have intelligence beyond the capacity one would have credited to so small and seemingly fragile a beast. If purposiveness can be accepted as an ingredient of intelligence, then the fox cub had it in good measure, as our later experiences of it showed.

When the cub had just been fed and was in a playful mood, it would retire behind the two-inch by two-inch wooden legs of the bench in the gypsy wagon. There it would look at me from behind the leg with one eye, then move its head and look from the other side with the other eye – exactly like an adult fox's trick of keeping you under observation while itself concealed. However, the cub's well-filled, barrel-like body protruding either side of the leg even when most of the rest of the body and head was hidden rather offset the effect!

I spent a great deal of time nursing the cub in my lap. After a while it would scramble up onto my shoulder and wander round the back of my neck, from shoulder to shoulder. It would jump from there onto the ground, landing perfectly on its four paws which looked too fragile to take the impact of such a fall. When travelling over the ground its gait was decidedly dog-like, lacking entirely the stealth of a cat. It would move in a direct line, usually towards an opening in a fence, when it was allowed out of the wagon, its front legs advancing alternately, its hindlegs being brought forward in a lolloping canter.

It was the second time I fed it that the cub began to show its paces. Having 'wolfed' the food handed to it bit by bit, there came the moment when it had eaten its fill. It took the next piece of food offered, ran to a corner of the wagon, dropped the food on the ground and came back for more. The next piece was taken to the same spot and dropped. This time the cub did not come back for more but lowered its head towards the food and, with hind-quarters hunched high, it bounced on its front paws in a way reminiscent of a puppy. After this, it picked up one piece of the food in its mouth and shook it, moving the head violently from side to side like a terrier worrying a rat. Then it trotted to another corner of the wagon, scraped away the peat litter with its front paws, placed the food on the bare board and shovelled peat over it with its muzzle. It fetched the second piece and buried it in the same way, after which it retired to sleep leaving two small hillocks of peat to mark its cache. In all this it was going through the motions of killing prey and cacheing it, something it had never seen done nor had itself done before. The following morning I found the hillocks gone, the peat levelled and the food eaten.

The cub was probably only four weeks old when it came to us and it must beyond doubt have been taken from its mother before its eyes opened. So the tendency to cache food is inborn since it could hardly have learned this from its parents. Moreover, we were to find other inborn tricks which would give us the chance to see which part of its behaviour was inherited and which was learned.

Although the cub slept through a large part of the day, and presumably the night also, there were periods of exercise when we tried tempting it to play. A ball was a moderately acceptable play-thing but its favourite toy was a piece of cloth or any similar object which could be seized by the teeth and shaken. Anything that could be tested with the teeth soon became ornamented with toothmarks and the highest mark of affection the cub could bestow on its human guardian was to take an outstretched finger in its teeth, without injuring it, then lick it with its velvet-smooth tongue. It would chew my shoelaces, wagging its tail all the time. The appealing look in its eyes, the manner of resting the chin on the forepaws, its fondness for being stroked under the throat and chin and the habit of resting the chin on your hand, were all reminiscent of a puppy. So, also, was its trick of rolling onto its back and opening its jaws in an obvious invitation to join in a

A cluster of aviaries, a few of those filling the garden at Horsley, that virtually drove the Burton family to find somewhere else to live

A view of Weston House as it appeared when we first inspected it

A tawny owl on the ground with wings spread sunbathing

One of the stone owls that stands sentinel on the brick pillars holding the gates at Weston House

A tawny owl in daylight, ready to be friendly and to greet us with bill clapping with its eyes sleepily half-open

ennie, our common genet from
Uganda, used human beings as
substitutes for trees, readily climbing
into one's shoulders, as seen here with
Jane

The lass with the delicate ears' sleepily
contemplating the camera. Genets are
related to mongooses but look more like
finely built tabby cats

Our pet genet making a long neck as
she stretches up vertically, her tail
acting as a counterpoise, to investigate
something just out of reach

Foxie, our tame fox cub, on the threshold of adulthood

The gypsy wagon which was not only a thing of beauty and antiquity but often served as a convenient nursery for young animals in need of care and attention

Foxie trying to take all the food into his mouth at once to carry to the vixen who has just given birth to a litter of cubs

Its first venture above ground: a fox cub being investigated by Foxie, the dog fox who sired it

Foxie with his vixen, Maxine, in the idyllic days of their courtship, engaged in a game of king-of-the-castle

rough-and-tumble biting match.

All this took place in early June. By mid August the cub had grown into a handsome dog-fox, complete with brush. He had become known as Foxie – not very original, but a name that had come naturally. Although he was completely tame he had been restricted to a large pen made into as nearly natural a habitat as possible. This was done by decorating it with stout boughs of evergreen, mainly conifer, so that the foliage remained green and did not need to be renewed too often. The sides of the pen were of stout chain-linked fencing, six feet high, and it was roofed with the same material, for a fox can surmount a six-foot wall or fence by jumping and scrabbling over the last foot or so. Chain-linked fencing was also buried under the floor of the pen, at a depth of a foot.

While endeavouring to retain his tameness, we treated him in such a way as to leave his natural impulses uninhibited. So we contrived, not without success I think, a compromise between retaining his affection as a pet and treating him as a subject for scientific observation, noting his activities and his reactions to changing circumstances and seeking to relate these to what we would learn of the ways of fully wild foxes.

Foxie's pen was forty feet from the house and twenty-five feet at its nearest point to the drive up from the front gate to the house. On the two sides facing the house and the drive the pen was hemmed in by aviaries, hurdles and a fair-sized garden shed. To a large extent, therefore, its occupant was screened from the human presence. Any stranger coming in by the front gate nevertheless drew forth the following behaviour. Foxie would jump onto the roof of his kennel and watch the stranger come up the drive. As the stranger drew level with the house he would jump down and watch him intently through the gaps between hurdles, shed and aviaries. Should the stranger now turn right and walk along the path running towards his pen, Foxie would scuttle quickly behind the foliage draping the inside of the pen or dive under his kennel, watching every movement of the newcomer while remaining unseen himself. It was an elaboration of his peeping from behind the bench leg in the gypsy wagon, when a cub. If one of us accompanied the newcomer, it was possible, with patience, to coax Foxie out of cover. Then, in due course, he might accept the stranger but it could take an hour or more, less perhaps if the stranger were a woman.

When I went alone to the pen, Foxie would watch me approach to within three or four yards and then bolt to cover. When I reached the pen he would emerge cautiously with ears, eyes and nostrils alert, then come forward a few paces and then bound back to cover again. In a few moments he would come out again, cautiously, to advance nearer to me before bounding back to cover. He might repeat this several times, coming nearer to me each time, until he finally reached the netting, to sniff my hand and lick it. This pattern was more or less invariable, even if I was speaking softly and encouragingly all the time.

Once he had identified me positively by smell I could enter the pen and there would be no further evasive action. He would offer his head to be stroked, behind the ears or under the throat, and lie down and roll over to have his underside stroked. He would stand on his hindlegs if I sat down, and rest his paws on my knee. He would allow me to pick him up, fondle him, stroke him, romp and play, running away only as part of the game. His favourite pastime was to undo my shoelaces with his teeth, and he seemed almost to take pride in being able to do this. He would rub round me like a cat, and when I left the pen, after a boisterous play period, the likelihood was that he would stand watching me go with an air of being disconsolate. Yet, were I to return a few minutes later, he would bolt for cover and go through the whole pattern of evasive action as before.

Jane took to feeding and tending Foxie. She also played with him more than anyone else. If he was attached to any one person more than another it was to her. Yet his reaction to her approach to the pen was little different to that with which he received me.

Early in his cubhood Foxie was introduced through the chain-linked netting of the pen to our great lumbering mastiff-like boxer-cross Jason. At their first meeting dog and fox touched noses, both waving their tails furiously in complete circles, but whereas the three-year-old dog maintained a stolid, and for him dignified, calm and silence, the cub was beside himself with ecstasy. Apart from his trilling and occasional feeble young cub's barks, Foxie had until then been quite silent. Now he trilled incessantly, licked the dog's muzzle, danced to and fro and from side to side and did his best to push his way through the chain-linked netting. After that, the mere sight of Jason, even in the distance, set Foxie off into his ecstatic display.

When Jason walked towards the pen there was no question of the fox retreating, no running for cover or going to earth. Instead, Foxie would station himself in the corner of the pen nearest the dog's line of approach, wagging his brush furiously, dancing in eagerness and trilling continuously. After a few such meetings we felt it safe to allow Jason into the pen. It was always the same performance: Jason would walk in with no sign of emotion, Foxie went nearly crazy with joy, the dog tolerating his ebullient greetings with sober nonchalance. If there was any food left in Foxie's bowl Jason promptly ate it, Foxie meanwhile crouching in front of the boxer-cross and looking at him in a manner that can only be described as admiring, as if it was an honour to be robbed by this huge companion.

So we had the pattern: at the dog's approach there was no retreat; at our approach the evasive action followed by the completely friendly reunion; at the approach of a human stranger the complete retreat to cover. There was, however, one important exception: if the human stranger were a child on its own there was no retreat, the fox showing pleasure but not so markedly as with Jason. Moreover, if a child, on its own, came through the front gate, Foxie would jump onto his kennel as usual and would jump down as usual as the child came up the drive and towards his pen, but only to advance as near to the netting as possible, all the time showing signs of welcome.

We saw this same recognition of juvenility in our tame rook, Corbie, whose aviary was nearer to the gate. If a stranger came through the gate the rook would retreat to the farthest corner of his aviary and show every sign of agitation. This became worse if the stranger approached the aviary, for the rook would then fly at the wire-netting in an effort to get away from the intruder. It would be quite some time before the bird would settle down and accept the person's presence. But if a child came in there was no such agitation.

Others of our animals showed this same tendency, not so markedly as the fox and the rook, but sufficiently noticeably to be significant. The difference in their attitude towards a juvenile and an adult must be influenced by impressions received through their senses. A rook's sight is keen, its sense of smell poor, its hearing mediocrely good. A fox's sight is probably relatively mediocre, its sense of smell acute and its hearing also acute. Yet there must be a common factor somewhere.

A theory which was much in favour at one time suggested that the parental instinct is evoked by the shortened face in relation to the large, bulging forehead, as in the human baby. We hear little of this theory today but it may be worth considering it in relation to this problem, since there is just a chance that the more benign attitude of Foxie and Corbie to juniors may have sprung from a parental instinct.

One day a boy of thirteen who was five feet ten inches tall came to see Foxie – who did not retreat from him. Another boy of the same height, aged fifteen, sent Foxie racing to cover. The only difference I could detect was that while the thirteen-year-old spoke in a piping treble, the voice of the fifteen-year-old had broken and was gruff. Both had the same proportions of face to forehead. Unfortunately for the theory of the shortened face, Foxie instantly made friends with Jason, who had a long muzzle; and he was instantly and violently antagonistic towards our black cat, in spite of its short muzzle and bulging forehead!

A short woman sent Foxie to cover, a tall boy did not – provided his voice was still treble. But a woman's voice is a treble. And can it really be that at a distance of thirty feet, in strong sunshine that makes its eyes contract to slits, a fox can be acutely aware of the proportions of a face, to make such a difference in its behaviour, especially since a teenage boy's face has much the same proportions as that of a young woman?

We often discussed this problem and reached no firm conclusion except to suggest that possibly both mammal and bird sense an absence of potential aggression in a boy and a woman, as compared with a man. In this context it may be recalled that certain hyaenas will, on occasion, attack women and children but not men, presumably recognizing the difference between them, and avoiding an encounter with the more aggressive man.

It is worth mentioning in this context that Robert, who was at boarding school and saw Foxie only in the holidays, with intervals of three-month's absence, found that the fox's attitude towards him changed when his voice had broken.

Perhaps the only safe conclusion to be drawn is that oversimplified explanations of behaviour should be regarded with suspicion, especially in dealing with organisms of advanced mental organization, such as foxes and human beings.

In November of 1956, when Foxie was about eighteen months old, a vixen was introduced to him. She was about the same age as he was. The introduction passed without incident. Indeed, far from showing animosity, Foxie grew wildly excited, rushing around the pen as the basket containing Maxine – she was given to us by Maxwell Knight – was carried towards the enclosure. Perhaps Maxine's scent recalled for Foxie the odour of his mother, from whom he had been parted before ever he could see her. At all events his excitement continued after he had gained his first sight of his new companion. Maxine, by contrast, appeared quite indifferent, which is probably explained by her sex – the coyness of the female – than from her being in strange surroundings. Gradually Foxie's excitement died down and when his supper was taken into the pen he attacked his food with concentration as if the nearest vixen was a hundred miles away.

So far as we were aware, the first excitement died down completely after the initial meeting and was not renewed for several days. What we did find was that Foxie continued to sleep in the earth he had dug for himself (after he had done this we had his kennel removed), while the vixen slept in the open. The second morning after her arrival the two were seen out in the open, he with his fur sleek, she with her fur frosted from exposure during the previous night. On the third day the vixen took to slipping quickly into the earth when the dog-fox came out for exercise and by the end of a week the two were sharing the earth. From this time onward the friendship developed slowly until, about two weeks later, they were seen playing together in a sustained and energetic manner.

Jane saw this first, her attention having been caught by noises such as jackdaws make. She went towards the pen and found the two foxes standing facing each other. The moment she appeared, however, they jumped apart. She waited quietly. The vixen ran back and forth across a log. Foxie went to sit in a corner of the pen. After a while he started digging over the straw covering the floor of the pen, running from one place to another with his ears laid back and making a clacking bark recalling the voice of a jackdaw. In her notes, made at the time, Jane recorded that a jackdaw flew over at that moment, so that she could confirm how similar the two sounds were.

Following the digging and running about, Foxie ran over to where Maxine stood on the log and they started what looked like a mock battle, each with the mouth wide open. Then suddenly each would stand up on hindlegs, facing the other, their paws on each other's shoulders, and for five to six seconds they shuffled their hindfeet in a sort of dance. The jackdaw-like bark could be heard throughout this time. A few minutes later Foxie again started to dig in the straw and to run hither and thither, then they sparred with open mouths again, and stood on their hindlegs with paws on shoulders. We saw this so often, this attitude of seeming embrace, in the years that followed, that we recognized it as the culmination of a set pattern of behaviour, but at that time Jane was so excited that she telephoned to London to tell me all about it.

For a number of weeks after this first *pas de deux* the play was a daily event. Only heavy or continuous rain tended to inhibit it. But on no single occasion was the play so vigorous and sustained as on that first day. I can be sure of this, even though I was not there to witness it, because as soon as Jane saw what was happening she rushed for her camera and filmed it. On this first occasion the two foxes ran through all their tricks and then repeated them. After that they repeated one or more but seldom all together and with so many repetitions.

The tricks, as I have called them, included the dog-fox running vigorously round the vixen while she lunged at him with open mouth; and the King-of-the-Castle game with one standing on a log and the other rushing around while both sparred at intervals with open mouths. Then there was a quieter game in which, again with open mouths, they slowly turned their heads from side to side. And, of course, there was the 'embrace', with paws on each other's shoulders. There were times when the dog-fox would drag the vixen over the ground by the scruff.

The initiative always seemed to be with the dog-fox. He usually started the playing and showed greater vigour at it. Sometimes he seemed to find it necessary to bully the vixen into taking part. One afternoon, for example, the two were lying curled up in opposite corners of the pen. Foxie began to play with a strip of felt, given him for that purpose. He worried it energetically, occasionally leaving it to rush over and pounce on the recumbent Maxine. Then he would take several turns round the pen at full speed, pounce on the felt again and shake it vigorously. He went back to Maxine again and they greeted each other with open mouths, she still lying down. First Foxie put his head on one side while Maxine held hers upright so that their mouths were almost interlocking, though not touching. If one moved its head so did the other, keeping the mouths in the same relative position. That time the vixen called, with a kind of whine.

Foxie now seized Maxine by the scruff and tried to haul her to her feet. She made no response as he tried several times to pull her up by the scruff or the fur of her flanks. He then pushed his nose under her head, pounced at her and nipped her paws with his teeth. She continued to lie on the ground with what may best be described as a languid expression. Finally, he pawed her face several times and took one of her ears in his mouth to pull her along. She gave a small cry of pain, jumped up and rushed onto her log. Foxie fairly hurled himself round and round her, rushing at her and dashing away again as Maxine countered his rushes with wide-open mouth, making a low guttural grunt each time she lunged at him. They played this King-of-the-Castle for five minutes, at the end of which time play suddenly ceased. Maxine went over and curled herself up tightly in her corner. Foxie, after sniffing around for a while, aimlessly investigating objects lying on the ground, followed her example in the opposite corner.

The following May Maxine bore a litter of four cubs. Before saying more about this let me digress. It was after the cubs of this litter were grown and had been dispersed to other homes that we took care of Charlie, a dog-fox belonging to a friend who had named him Charles James Fox after the well-known English statesman. While Charlie was with us, he was housed in a separate pen a few yards from the pair. He was larger than Foxie and more handsome.

One day, Jane decided to let Charlie into the pen in which Foxie and Maxine lived, so that he could enjoy a little company. When Charlie entered the pen the first thing he did was to drive Foxie into cover and, throughout Charlie's stay of about an hour, all we could see of Foxie was one eye peeping through a screen of foliage. With Maxine it was different. She fairly squealed with delight and made a great fuss of Charlie. I have an indelible memory picture of Maxine at one moment lying on her back in front of Charlie, kicking her legs in the air and squealing, squealing, squealing in apparent ecstasy. Charlie showed less emotion but when, a few moments later, Maxine got onto her feet Charlie mounted her and tried repeatedly but ineffectually to mate with her. This was remarkable since it was nowhere near the mating season. Foxie was forgotten, so far as Maxine was concerned, and he made no attempt to emerge from cover to assert his marital rights. It was the clearest instance of animal cuckoldry that one could hope to see.

Charlie was later returned to his own pen and was not allowed in with Maxine and Foxie again. They continued to live together for another five years, Maxine producing one litter after another, usually of four cubs, except for the spring immediately following the encounter with Charlie. The whole history of Foxie and Maxine, with the brief interjection of Charlie, reminds me forcibly of the human situation that sometimes occurs when a woman falls passionately in love with a man, whom she is unable to marry for one reason or another. She is, however, forced by circumstances to marry another, bears him a family, serves him devotedly as wife and companion, yet eventually, with her dying breath, whispers the name of the true lover from whom she had been forever parted. This episode, as between Maxine, Charlie and Foxie is, for me, the most clearly delineated instance of many I have met which suggest that some animals do choose their partners, if allowed to do so. That is, they fall in love as truly as any humans, indicating that the mere gratification of a sexual urge is not the whole story.

More recently, the results of observations on the domestic dog, published in *Animal Behaviour*, indicate that a bitch kept with several dogs will favour one of them as a companion, but prefer another as a suitor.

Foxie, by contrast, was faithful to the end, a model father to his many cubs and enduringly devoted to Maxine. In his devotion to her we learned much, and especially about the remarkable change in his behaviour as the birth of the cubs drew near.

We did not witness the mating and it was not until a month before the cubs were born that the vixen showed any unusual behaviour. Then, one night, about an hour after nightfall, she cried out three times, each time uttering a trio of screams, more like a child in distress than the usual fox scream. The next night she screamed again, in the same manner but for a longer period; and the same on the third night. Again, a week later she screamed and continued almost throughout the night. After that we heard nothing more from her.

Soon after Maxine first came to us we had made her an earth, with a single entrance, but she did not sleep in it until she had herself made a second opening to it, a bolt-hole presumably, in case of trouble. Shortly after the screaming period she was seen to be making a tunnel which started in front of the original entrance to her earth and ran for three feet. She had only a foot of soil in which to excavate it, because of the chain-linked netting buried in the floor of the pen. Then she blocked up the exit from the earth. She now had her sleeping chamber with its single entrance and, immediately in front of this, the opening into a tunnel which had another opening to the surface at the other end.

By now, the vixen was showing in her girth the events we should have expected. She was also stripping the fur from her underparts, not only exposing her teats but removing the fur from the insides of her legs as well. Throughout these operations the dog-fox and the vixen maintained an impersonal attitude towards each other, although Foxie did assist a little with the digging.

From the time Maxime first arrived Foxie had been greedy. We found it necessary to put the vixen's food on a separate plate set well away from his or she would have starved. Therefore, her food was now placed near the outer end of the new tunnel, in which she was spending more and more time. This was in the hope of defeating the characteristic greed of the dog-fox, which I would emphasize again because of what followed.

There came the day when the plate of food was placed near the mouth of what we later called the nursery earth. Foxie came forward, pressed each piece of food with the tip of his snout, in the way he would press food into the earth when cacheing it. Then he covered the whole with straw and left it. What happened afterwards we do not know, except that by the following morning the straw had been removed and all the food was gone.

The following evening, when the food was taken in, Foxie took a piece delicately in his front teeth and made for the entrance to the nursery earth, uttering a low bark we had not heard him make before. The vixen came out of the earth, took the food from him, ate it, then herself went to the plate. So, night after night, he continued to take food to her before partaking of it himself. This was in marked contrast to his behaviour previously, when he had wolfed everything he could without regard to her needs. Selfish to a degree, he had suddenly become unselfish.

The cubs were born on 19th May. For several days before and again after that date, the vixen left the earth, but this was rare. When she did there was no rough play. Instead, it was the vixen that more readily approached the dog-fox: she would go quietly up to him, as he was sitting on his haunches, and she would sit closely beside him in the same attitude. From time to time the two would turn their muzzles towards each other so that only the tips met and each in turn would gently lick the other's lips. The whole action had an air of gentleness and devotion. All this and everything else in the development of the cubs was recorded on film, the whole sequence taking a full hour to project.

At times, when the two were sitting side by side nuzzling each other, the vixen would close her eyes as if the touch of her mate's muzzle was excruciatingly enjoyable. What we then had was a picture of perfect marital bliss.

A week after the cubs were born a small part of the roof of the nursery earth fell in. It was then that I had the first firm evidence, which I had been waiting for, of the high mental standard of the species. Maxine showed no panic, although the suddenness of the fall must have surprised her. She calmly left the nursery earth, picked up one lump of soil after another and deposited them clear of the tunnel; then, equally calmly and without haste, she took each of the four cubs in turn and carried them into Foxie's earth. He allowed her occupation of his private apartment without visible sign of protest.

It was noticeable that, immediately after this change of sleeping quarters, Foxie at feeding time would take food towards the now-derelict tunnel, hesitate a moment at the gaping entrance, then move round by a circuitous route to where the vixen was in residence. Sometimes he seemed to be bewildered and when halfway there would hesitate, as if he was not sure what he was supposed to be doing; then he would swallow the food or else cache it. This period of indecision was, however, brief.

The dog-fox's earth, now taken over by the vixen, also had two openings, in line with each other, so that sufficient light entered to allow us to see its occupants dimly from a few feet away. We were anxious not to disturb the new mother, so we kept our distance. It was possible, even so, to see the sightless cubs, with their rounded muzzles and short tails, looking very unfoxlike. We could also see the vixen lift each in turn with her mouth and place it in position, so that when she lay down, or turned round within the earth, she neither injured them nor denied them the protection of her own body. It was a delightful picture of maternal care, devotion and solicitude.

Careful as we were, however, even this intrusion was too much for her, and in a short time, while nobody was looking, she transferred her litter to a third earth, one Foxie had dug for himself. So far as we could see, he surrendered his apartment to her yet again, without demur.

Eight years after Foxie first came to us, Maxine was again in cub and birth was imminent. At such times, apart from the birth of the first litter, when our curiosity was at its height, we disturbed her as little as possible. We knew she had retired to her nursery earth as usual so we kept clear of the pen, except to put in food twice a day. That evening we found Foxie free in the garden. At first he allowed us to go near him, but not near enough to take hold of him. We hoped he might return to the pen of his own accord. This he failed to do. Every now and then he would peep out of one of the shrubberies and as time wore on and the daylight was fading he allowed us to approach less and less closely to him before bolting into cover.

Meanwhile, we inspected the pen. Maxine lay dead in the nursery earth, on her side, her belly swollen with the cubs that would never see the light of day. We found one of our bantam hens pressed into her body. Presumably Foxie had sensed something was wrong and had tried to help in the only way he could, by fetching her food. He had forced apart the stout links of the pen's chain-linked fencing with his teeth and paws, and had struggled through an incredibly small opening. He had killed a bantam hen and dragged it and himself back through that small opening. Presumably, when Maxine did not respond to his gift-giving he had pressed the dead hen tightly into her body, under her chin and between her forelegs, as she lay dead. Then he had left the pen again by the same hole.

I was up at first light the following morning scouring the garden for a sign of Foxie. At about seven o'clock someone going to work knocked at the door to say he had seen a fox in a nearby field and to ask if we had lost one. I went out to look for Foxie but could not find him. Later that morning somebody else called to say that our fox was now in another field playing with the lambs.

I knew I would never be able to catch Foxie; he had gone wild. I felt certain he would not harm the lambs, unless he grew hungry. In any event, if he did kill a lamb the guns would be out after him. With a heavy heart I took the quick decision to phone the gamekeeper, explain the position and ask him to shoot him. Within a few minutes I heard two shots ring out. I could not bear to go and look and I was grateful that the keeper had given him both barrels. If this gallant little gentleman had to go, it was best he should die instantly.

5 *Corbie, the rook*

Acquiring animals is like any other form of collecting, even when one embarks on it with the highest of motives, which in our case was to give sanctuary. The appetite grows by what it feeds on – and one soon becomes insatiable. When we learned that somebody in Wiltshire had a tame rook for which she needed to find a new home, therefore, we readily agreed to adopt it, although it meant a long journey to fetch it.

Diana Ross, the novelist, had picked up the rook as a fledgling from a gutter in the city of Bath, while she was shopping. It was muddy and dishevelled and she had taken it home and hand-reared it. Corbie, as she had named it, was now a fully-grown male. He had the freedom of her home in Melksham, and although he spent most of his time in the house and the surrounding garden he not infrequently spent the day in the fields with the wild rooks. He always returned home to roost in a parrot's cage suspended from the ceiling in the kitchen. Every evening, on the command 'Up', he would fly onto Diana's shoulder and from there into the cage.

Corbie could have been sent to us but Diana Ross wanted us to see all his tricks for ourselves and she was afraid he might, on being transferred to a new home, lose them. In fact, this did not happen but I was very glad we made the journey to fetch him. He not only understood the command 'Up', for example, but immediately responded to 'Out' when ordered out of a room. There was, indeed, complete understanding between owner and pet: not only did he respond promptly to her words, but Diana had done her best to learn rook language! The most bizarre event, one of many during our visit, was to see Corbie on her shoulder, she rubbing his plumage with the side of her face, he nestling to her, and all the time the two conversing in the throaty gurglings that constitute the intimate language of rooks.

The rook's prying habits had led him to peck open a box of self-striking matches. Then, attracted perhaps by their red heads, he had held one match after another in his foot and pecked at the head making it burst into flame. As each match came alight, he picked it up and held it under his wing, spreading the wing for this purpose. Diana had been obliged to buy only safety matches, but she had a box of self-striking matches for our visit, to let us see Corbie in action.

One of the first things we saw, on entering the house, was an open grate with a fire. We saw Corbie walk towards the fire and we heard Diana say 'Out', whereupon Corbie turned and strutted out of the room. Diana explained that she had had to buy a fireguard because Corbie had walked onto the burning logs several times, and she was afraid he might be burned to death. It was when we were in the dining-room that I noticed Corbie, in front of an electric fire, holding his wings in the typical 'anting' fashion. The fire was not switched on. I drew Diana's attention to this with the words: 'That bird is anting.' She replied 'I know.' Up to this moment nothing had been said on the subject, and the word 'anting' had not even been mentioned. It transpired that this was what Diana had wanted to show us, what she had meant in her letter by Corbie's 'tricks'.

Anting has been observed in some 250 species, all belonging to the order *Passeriformes* or perching birds. Briefly, when a bird ants it picks up an ant in its bill, raises its wings – in some species only one wing is raised at a time – and rubs the ant along the underside of the flight feathers. The wings are only half-spread and they are curved in a special way that once seen cannot be mistaken for anything else. At the same time the tail is twisted to the side on which the ant is being rubbed. The ant may then be swallowed or cast aside and usually another ant is picked up and rubbed on the other wing. The posture adopted, when a bird is anting, looks somewhat grotesque but there is something like ecstasy in the bird's movements. The pattern varies slightly from species to species, but once you have seen one bird anting you would never fail to recognize it in another species.

What is puzzling is that birds can be seen 'anting', that is assuming the posture and making the same movements, in smoke from chimneys or from burning logs. A bird may ant when merely standing on ground where ants have been running and, exceptionally, birds have been seen to ant with such odd things as moth balls, lemon juice, apple peel, vinegar, hot ashes and a variety of other substances.

All this had become known to ornithologists between 1934 and the time we agreed to give Corbie a home. What we now found was that we had a bird that would only indulge in the typical anting posturings in the presence of heat.

Meanwhile, Diana went on to tell us how she had come back from a shopping expedition soon after Corbie was mature, to find the electric fire switched on. She was confident it had been switched off when she left the house. Then she discovered that Corbie had learned to switch it on. This alone is remarkable because the fire itself, standing against the dining-room wall, was a full three feet from the switch. So either the rook had seen her switch on the fire and had copied the action, or in his moving around pecking at random at everything that aroused his curiosity – which is a marked feature of rook behaviour – he might have struck the switch with his bill, discovered the heat from the fire and thereafter switched the fire on whenever he felt so inclined Whichever way it was, it showed unusual perspicacity on the part of the rook.

The rest of the story is worth telling. The next time Diana had to go out, having once found the fire switched on, she turned it to the wall. When she returned she found the fire had been switched on again, and the wall was scorched. Always, after that, she made a point of pulling the plug from the socket when the fire was not actually in use.

Another of the rook's tricks was demonstrated fortuitously a minute or two later. Corbie, as I have explained, had the freedom of the house. All through tea he was wandering over the table, picking his way carefully among the plates and cups. The teapot was dark brown, its lid pale yellow, and Diana explained to us then that Corbie had broken one lid after another by hammering at them with his beak. He did this to ant in the steam rising from inside the teapot. While we were there, in fact, he knocked the lid off the electric kettle and anted in the steam.

It was after tea was finished that we were led into the kitchen to see what Corbie did with a box of self-striking matches. We saw him hold a match in his foot on the stone-flagged floor and ant with one burning match after another. Having watched this performance we stood in the middle of the room discussing its implications. Then I noticed Corbie was perched on the hot tap over the kitchen sink. He was pecking vigorously at the tap between his toes. Diana explained that he would ant in the steam from this tap and was forever trying to turn the tap on for himself, to enjoy the steam. To demonstrate she turned the tap on so that we could see Corbie in action.

Incidentally, Corbie seemed to have a penchant for dealing with mechanical devices. Soon after he was finally installed in his permanent aviary we found, time after time, that the door to the aviary was open and Corbie was walking around the garden. By watching him we found he had discovered how to latch and unlatch the cabin-hook that fastened the door on the outside, so a rook-proof fastening had to be devised.

To begin with, however, he had to be placed in a temporary cage, in solitary confinement. While he had been at Melksham, Corbie had had as boon companions, apart from Diana, a she-cat and a domestic hen. Not surprisingly, for three days after he came to us he moped; he was missing his friends. We kept watch on him anxiously. Late on the third day I happened to be looking at him, sitting hunched up on his perch, when suddenly he straightened himself, made the noise of a domestic hen that had just laid an egg, and from that moment he cheered up and was quite normal, although still in solitary confinement.

Once Corbie was installed in a permanent aviary I began to supply him with self-striking matches. These were available in two different brands sold in boxes of markedly different sizes. The smaller kind he had known at Melksham. The larger kind he had never seen before, so far as I could ascertain. Yet the first time I took this larger box out of my pocket he came over and solicited it, although its colour and the pattern of the label as well as the shape and size of the box were different from those of the smaller box. It seemed that to Corbie a matchbox was a matchbox whatever its differences from the one he had been used to.

I also tried him with lighted cigarettes and he treated these as he did lighted matches: he rubbed the hot end first under one wing, then the other. If a cigarette had been lighted but had gone out he would take it and rub the cold ash under his wing. He would do so more vigorously if the ash was still warm. When a lighted cigarette was held so that the smoke drifted into the aviary he would peck at the smoke wisps and ant with imaginary beakfuls of smoke. He would do the same with smoke from any source, from smouldering rag or paper. Always, and I must stress this, he adopted the same posture and made precisely the same movements as when anting with ants. If, therefore, this behaviour cannot be included as true anting, then we really have a puzzle: why should the bird be so obsessed with heat, smoke and ashes and treat them as he would ants?

Ornithologists seem to have concluded finally that anting is a matter of feather maintenance, the assumption being that the formic acid from the ant tones the feathers or kills feather mites, or both. It is a conclusion I find a little doubtful for several reasons which will emerge later. A writer in the *New Dictionary of Birds* even claims that the smoke bathing, already foreshadowed here in Corbie's behaviour, should be included under feather maintenance. But it is hard to believe that smoke can have any beneficial effect on feathers.

What I am more concerned with here, however, is the evident obsessiveness. In those early days, if I approached his aviary for example, Corbie might 'talk' to me or display to me but he would not necessarily fly over to me. If, on the other hand, I was smoking he would come across the aviary with what can only be described as eagerness or even excitement. He would land on the perch nearest me, put his beak through the wire and virtually ask for the cigarette. Corbie would leave his food, even when he had not fed for a long time, to take a lighted match or a burning cigarette from me. He would go half into the anting posture at the sight of an ant or a match, a matchbox or a cigarette, held as much as twenty feet from him. I kept my cigarettes in a tin box in my hip pocket and once he had seen me take out the box to extract a cigarette he would get excited, even going partly into the anting posture, at the mere sight of the tin box. Later, when he was more used to me, I merely had to put my hand towards the pocket, as if to take out the box, for him to get excited.

It is hard to believe that all this excitement and obsession were in pursuit of what should be an ordinary commonplace function such as feather maintenance. If I lit a wisp of straw in his aviary he would show the same excitement, fly down and seize a piece of glowing straw in his beak. It is even harder to see how glowing embers could benefit the feathers. Rather the contrary, for although we had several birds at different times that would ant with lighted cigarettes or burning straw, and although none ever did itself significant damage, occasionally we could smell a slight odour of scorched feathers after they had performed with burning materials. This is hardly the way to tone the feathers, and since nobody has produced any irrefutable evidence that formic acid tones the feathers or kills destructive parasites, such as feather mites, the notion of feather maintenance rests on an insecure basis. It was demonstrated by a Russian zoologist that mites are killed when a bird ants, but the mites he named were those that are beneficial to a bird in keeping the feathers clean.

In 1955 I recorded that anting birds always start with the left wing. Ten years later, when we had several anting birds in our aviaries, I made observations on this over several weeks. Each day during that period I induced each of these birds to ant, with ants or lighted cigarettes, and I noted which wings were used. Invariably, each bird started with the left wing each time. Afterwards it might then put an ant or the lighted cigarette under either the left wing or the right wing, and usually it went on to ant several times in quick succession.

All this I tabulated and when, at the end of the time, I analysed my tables it was clear that, on average, birds ant three times under the left wing for every once under the right. If anting results either in toning the feathers or destroying harmful parasites, then habitual anters, of which there are plenty both in the wild and in aviaries, ought to have noticeably better feathers on the left than on the right side. There is no evidence that this is so.

There was an occasion, for example, when, coming round a garden shed, I saw fourteen young starlings on the ground, about six feet from me. At that moment they started to ant, each under the left wing. Then they became aware of me and, as starlings will, all took off simultaneously and flew up into a nearby tree. Because the starlings were strung out on the ground almost in a line – and because one develops the habit of counting numbers as part of one's natural history observations – it was easy to count them and also to see which wing was being used. Indeed, the impression I retained of the starlings was of something very like a drill squad.

While the starlings were in the tree I took the opportunity of counting them again. At the same time I remained absolutely immobile, which is what any experienced naturalist trains himself to do, hoping they would repeat the performance. Sure enough they did. They came down to the same spot, again landed almost in a straight line, and all, more or less simultaneously, anted again, each with the left wing. This time, however, most of them anted a second time, all again using the left wing except for one that used the right wing.

This sort of observation occurs rarely, but it seemed almost a textbook example of the preponderance of the left wing in anting. It is interesting, too, to recall my visit to H. Roy Ivor, at his sanctuary in Windinglane, Toronto, Canada. I was taken there by Professor Milner and his wife Jean. During the visit, Roy, who had been studying anting for many years, invited us to see his numerous transparencies of this phenomenon. He projected dozens of these photographs onto the screen and towards the end Jean suddenly said: 'Why do they all use the left wing?' She had noticed something which Roy Ivor, with all his experience, had failed to see.

Another feature of anting is its apparent infectiousness. Several observers have mentioned this. We noted on several occasions how, when one of our birds had been induced to ant, either with ants or with lighted matches, all the birds in adjacent aviaries – if they could see what was happening – grew perceptibly excited, or went into a partial anting posture.

There was a Sunday morning when a naturalist friend called on us to be shown our birds anting. I had obtained some ants from another part of the garden and put them in Corbie's aviary. He immediately started anting vigorously, almost as if in a frenzy. At the same time every member of the crow family in the aviaries around went into the anting posture but without having any ants to pick up. There were two jays, two crows, two rooks and two magpies; most of them were males, but some were females. The sight of all these birds simultaneously in grotesque anting postures was strikingly bizarre, to the point of being weird and uncanny, and not easily to be forgotten. I distinctly recall one magpie flying at the wire-netting of its aviary as if trying to reach the ants, hanging on the netting by its toes and at the same time going into the anting posture, even putting its beak under its wings.

This scene is linked in my memory with a similar event two weeks later. I had gone into one of the aviaries and the crow in it landed on my head. I could not see it but I felt it crouching and flapping its wings and could hear it making the calls that accompany copulation. Immediately, in the aviaries all around me, the other crow, the jays, rooks and magpies, each where it was at that moment, on the ground or on a perch, went into the attitude of coition and uttered the guttural copulatory noises appropriate to its species and sex. It lasted only a moment or two, but the effect was striking, weird, even frightening – as if the crow on my head had unleashed a demon.

It is events of this kind that make me feel that my fellow scientists have wandered along the wrong path in concluding that anting is no more than a matter of feather maintenance. Over a period of almost twenty years there were, at one time or another in our aviaries, nearly twenty members of the crow family. Some were habitual anters, others never anted to my knowledge. The habitual anters only anted consistently during the few years I was studying the subject, using ants, matches, cigarettes, flames or smoke which I made available to them experimentally. Throughout that time they all had marvellously glossy plumage, as Jane's many films and photographs of them show. The secret of good feather maintenance lies in the food. Well-fed birds, as ours were, are healthy and their plumage is maintained in perfect condition. They had no need of anting for that purpose, yet some of them were obsessive anters.

It would be wrong to leave the impression that our only interest in Corbie lay in his anting. He taught us much more than that. When he first came to live with us, for example, he quickly began to show preferences for certain people. Noticeably, he was much more favourably disposed towards women than men. This might be explained by his having been rescued and hand-fed by a woman, so that we could say he had become not merely human-fixated but woman-fixated. It was also noticeable that he not only showed a greater preference for Jane than for any of the rest of us, but frequently favoured her with the courtship characteristic of his kind. That again might be explained by the fact that it was she who always took him his food. Even so, it was quite astonishing, to me at least, to see the way he would fly to her, stand on her hand and go through a most elaborate courtship display, occasionally ending by depositing his semen in her hand. To all intents he treated her as another rook – fixation could hardly go further.

In due course we acquired two young crows. Jane had found them starving and in poor condition, in a field. We eventually let them have their liberty but they made such persistent efforts to get into Corbie's aviary that, for the safety of all three, we had to allow them. Very soon Corbie 'adopted' one of them. He began to make affectionate sounds towards her and, as soon as food was put into the aviary, took some in his beak and fed her. She responded in the usual way, by crouching, spreading her wings and fluttering them, and holding her beak wide open for the food. Now Jane found that she entered Corbie's aviary at her peril: there was always the chance he would fly at her and lunge at her with his sharp beak.

The other crow had to be removed, leaving Corbie and his chosen companion, whom he courted continuously throughout the rest of the year. He did so to the accompaniment of a delightfully musical call, a somewhat metallic sound that can best be rendered in print by the word 'clock'. Even when we were not able to see the aviary we could always tell what he was doing by hearing this call. He was sidling up to the crow with his wings spread at his side, the flight quills fully extended and the tail feathers spread fanwise. Rooks and crows are always described as black but close to their plumage is seen to have a purple or blue iridescence and a rook in full courtship display presents a magnificent sight, especially in strong sunlight.

Corbie also began to show an interest in any small stick or twig he found on the ground. He did no more than pick it up, carry it around, then drop it elsewhere. At the beginning of the next year, in January, his interest in sticks increased and Jane took a supply of them into the aviary and scattered them over the ground.

Simultaneously another feature of his behaviour became marked, too marked for the comfort of anyone entering the aviary. Instead of being hand-tame as before, he tended to show aggression towards any intruder into the aviary. He also took more and more to a particular perch, one of two set at an angle, and when there his belligerence was noticeably increased. Whether he was perched there or not, anyone going near that perch or putting a hand to it was violently attacked.

It soon became clear he was going to build a nest, but would find this difficult with only these two perches, so Jane put a stout stick in position to assist his efforts. She retired hastily from the aviary after having done so with several scratches down the side of her face. At all events, he now had three perches arranged in a triangle. With only the original two perches, set at right angles to each other, he had optimistically started the foundations of a nest. Even with the third stick added, to complete the triangle, the sticks he placed in position did not always stay there, but in time a firm base was completed. Then more sticks were added, and a lining of dried grass completed the task.

When the nest was in its final stages, Corbie tried to introduce the crow to it. The breeding period of a crow is later than that of a rook, which may account for her reluctance to be involved, although she seemed to be watching what was going on with interest. She made no effort to assist, although all the time Corbie continued to pay her court and she to accept gifts of food from him. He did, however, manage to coerce her onto the nest from time to time. Choosing a moment when she was perched near it, he would advance towards her, lower his head under her breast and force her backwards until, in the natural course of retreat, she stepped backwards into the nest. Then he was content, even if she immediately stepped out of it again.

Nest-building is inborn, all members of a species building to the same pattern. Nevertheless, some learning is involved. From carefully watching Corbie it was clear he was working to a design, the unfolding of which must be innate. At the same time, it also seemed evident that as he progressed his skill increased. That is, something more than purely automatic behaviour was involved. In the early stages of building, for example, he would carry a stick up to the nest, place it in position, look about him, return to that stick, pick it up and place it elsewhere. He might do this a dozen times with one stick. As time went on, however, he placed his sticks with greater assurance and was less inclined to move them once they had been positioned. It appeared that he was able to judge from the size and shape of the stick where it should be placed to the best advantage. In selecting the sticks from among the heaps we had thrown onto the floor of the aviary, his actions also passed from a marked indecision in choice to a perceptible confidence.

Perhaps the clearest instance of positive learning came in his handling of one particular stick, an irregular Y and about a foot long. To carry each stick up to the nest site, the rook would fly up to a low perch, then onto one at right-angles to it but above it, onto a third perch, then onto a fourth, and so to the nest. At each step there were hazards. At the first perch the awkward Y-stick became jammed and Corbie was flung backwards, releasing the stick, which dropped to the ground. There it lay until the next day. Again he tried and this time, profiting by experience, he held the stick so that he landed successfully on the first perch. When he flew to the second perch, however, the end of the stick became caught in the wire mesh of the aviary. Again the frustrated Corbie had to let it fall. In the end, he took nearly a week to carry the stick up to the nest, with intervals of several hours, or sometimes a whole day, between successive attempts. At each repetition, it was very clear from his manoeuvres that he was remembering previous experiences and avoiding the obstacles encountered in previous attempts. In the last attempt but one, the rook had landed on the perch normally, and, as he turned his head to take off into the nest, the end of the stick went through the wire ceiling of the aviary and was whipped from his beak. This time he flew immediately down after it and when he reached this same stage of the journey on his return, he lowered his head before turning it and flew successfully onto the nest. There, without hesitation, he placed the stick in position, patted it with his beak and left it. It was the sort of pat a woman gives to her dress when she has arranged it to her satisfaction.

These may be trivial details, and they represent only a brief sketch of all that took place. They do suggest, nevertheless, an increasing skill in an operation carried through for the first time; and – dare we say it? – a slight modicum of reasoning in its performance.

The result of this activity was that Corbie had built a nest very like any other rook's nest but, because everything took place within an aviary, at no point higher than nine feet from the ground and in full view at all times from the front of our house, we could keep a constant watch on events. In some ways we were watching an unnatural procedure. In the wild, rooks nest high up in tall trees. They bring in sticks, with which the body of the nest is built, by flying high and descending onto the nest as they land. In an aviary, even a roomy one, the sticks must be carried up by the bird half-flying, half-hopping, from one to the other of several perches. This presents hazards that are not normally encountered in the wild, or not encountered to the same extent. It may therefore not have been without significance that, in this first year, Corbie showed a perceptible improvement, as nest-building proceeded, in negotiating obstacles while carrying large sticks.

The next year Corbie built another nest. In the interval he had pulled the old nest to pieces, so that although he chose the same site, it was necessary for him to start from the beginning again. This time there could be no question but that the skill learned, and practised, last year had been retained. The whole building was carried out in markedly less time. In taking sticks up to the nesting-site there was incomparably greater skill in negotiating obstacles, and noticeably greater confidence in placing the sticks in position with far fewer changes in position after a stick had been placed. Above all, there was a most decisive air in selecting the sticks.

The previous year, he had approached a heap of sticks, picked up one after another and thrown them over his shoulder. Then he would go over them again, picking them up one at a time, testing each in his beak, rejecting and picking it up again, as if unable to make up his mind. This year it was as if he knew which stick he wanted to use before he took it in his beak.

Perhaps the most surprising feature in this nest-building activity was the part played by Corbie's companion, the hen crow. It has been observed, in the wild, that it is the male in both rook and crow that builds the nest. The usual statement in books on the subject is that the hen assists. It is true that the association of rook and crow is unnatural, so that behaviour could be distorted. However, this particular hen showed an interest in the nest-building and appeared to be helping – until one watched closely. Then it was apparent that often the reverse was true. After Corbie had carefully placed dried grass in position to line the nest, she would inspect it and, as often as not, tear it out and carry it off. Then Corbie would follow her, much ruffled, uttering low, discordant notes, snatch the grass from her and, with every appearance of high dudgeon, return to the nest and replace it. The purists tell us that anger is a human emotion, and that animals are not capable of it; but Corbie showed every sympton of anger on such occasions, which seems to me very nearly the same thing as being angry.

When the nest was completed the crow laid two eggs. We were quite excited about this and debated whether the hatchlings should be called rows or crooks. Unfortunately, the crow suffered a prolapse of the oviduct and, in spite of the ministrations of the vet, was found dead in the nest a few days later. For two days, Corbie showed every sign of grief. He remained perched in one spot, near the nest, his head sunk in his shoulders, looking the picture of misery. Food seemed not to interest him. Then, at the end of two days, he suddenly became normal again. If this was not grief, in its true sense, then I have yet to see it.

That was not the only sad outcome to the story of Corbie and his mate. The other was that it did not occur to me until too late to put the hybrid eggs into an incubator. We were all much moved by the sudden loss of a pet, just as Corbie was upset by the loss of his mate, and this deflected our thoughts from more practical issues. Whenever I recall this episode I regret my stupidity in not thinking of an incubator.

6 Ermintrude, the stoat – and others

Ermintrude, or Ermie for short, was given us by Nicholas Tindall, a young man who lived in the nearby city of Guildford. One day, at school, walking across the playground with a companion, Nicholas saw a stoat with something in its mouth. Thinking it was a small bird the stoat had captured, the two boys gave chase. The 'bird' proved to be a baby stoat. Its mother chattered at the boys from the long grass but seemed prepared to lose her offspring rather than take risks with her own skin, so Nicholas was left with a baby to feed.

Nicholas took the baby stoat home to hand-rear it. His mother had mixed feelings when Nicholas arrived with his new-found pet, but the sight of the baby, and of Nicholas' eager face, soon overcame any prejudices she might have had, and the young stoat joined the Tindall menage.

All went well until the stoat began to reach full size. The Tindalls lived in a flat, which was the first disadvantage. Then Nicholas began to neglect his pet and Mrs Tindall found herself looking after it, so she issued her ultimatum – that was how Nicholas got in touch with us and asked us if we would take Ermie.

When Nicholas arrived he was invited into the living-room and as he stood before us the stoat climbed over his shoulders and dived into his pockets, all at the usual incredible speed of these small, long-bodied carnivores. At one point, in clambering over the young man, Ermie paused to lick his lips in what must pass for a kiss of affection. Finally, she took a flying leap and, landing on my shoulder, gently bit the lobe of my ear but without drawing blood. Clearly, she was a very tame animal.

One reason why we were glad to take the stoat was that she would give us yet another subject for study. Naturally, the cameras were always at the ready to record anything interesting she might do. The photographic results were, however, disappointing largely because she usually chose to perform her most interesting tricks when she was out of range of a camera – and because of the lightning-like speed at which she moved. From a purely zoological point of view the results were more rewarding. If we learned nothing else we became fully instructed on just how quickly a stoat can move when it chooses, and on the agility it can display in rapid manoeuvre and also in climbing, provided it has the slightest hold for its claws.

The outstanding feature of Ermie's activities, as we saw them, was the intensity with which she would play with a particular object and then how quickly she would tire of it. We noticed this first on the evening after her arrival. The film camera had been set up in front of a deep armchair. Ermie was placed on the seat of the chair and, practically without hesitation, she set out on a voyage of exploration. There was a swift flash of light brown as she disappeared over the arm only to appear a second later upon the top of the chair's back, peeping from behind a cushion resting on it. From there she travelled rapidly over the sides, back, arms and every other part of the chair, ending up with a game of hide-and-seek in which she appeared from under the cushion, first on one side, then on the other.

Tiring of this she leapt onto an occasional table on which stood a flowerpot containing a treasured plant. In no time at all, Ermie had conceived an ambition. It was to stand on her hindfeet and, with lightning-like digging actions of her front paws, empty the pot of its earth. It says much for my wife's forbearance, in this and all other of our activities, that she stood and watched her precious plant on the way to being uprooted with the utmost calm – at least outwardly. I had long been aware that stoats climb well and have been seen high up in trees. I had not, however, credited them with being accomplished diggers.

Ermie was attracted away from the pot-plant with a woollen sock, which Jane snatched from the mending-basket nearby and dangled before her nose, enticing her back onto the armchair. The sock was then dropped onto the seat for the stoat to play with. She seized it in her teeth and started to roll with it, her long lithe body twisting, turning and somersaulting, all at high speed. In the end stoat and sock were hopelessly intertwined. They looked like a pair of misshapen snakes engaged in a vigorous wrestling match. The wrestling was alternated with the stoat holding the sock in her teeth and dragging it backwards. It was noticeable that, almost invariably, Ermie held the sock by the toe end, the equivalent to her, presumably, of the head end of a prey animal. Another plaything she had was a knitted woollen teapot cosy, with a pompom tassel. We noticed that in holding this she always gripped the tassel with her teeth, which was an even closer approximation to the head of living prey.

At one point in her performance with the sock Ermie left it lying on the seat of the chair and performed a circus act over it. She leapt swiftly from one arm of the chair to the other and back again, repeatedly and in rapid succession, bounced onto the sock itself, ran around the seat of the chair, back onto the arms, and then repeated the whole sequence many times, always at incredibly high speed.

Suddenly tiring of this, the stoat jumped to the floor and flashed across to what we called the Italian chest, an antique settle, its seat six feet long, its back nearly as high, and the whole heavily carved and copiously inlaid. It stands against a wall with less than an inch between back and wall. Behind this handsome piece of furniture Ermie was thoroughly at home. We could hear her climbing up and down and back and forth behind the settle, every now and then peeping out with a quick thrust of her head, which was as rapidly withdrawn. Nothing would tempt her out, so we were compelled to drag this heavy piece of furniture forward in the vain hope of catching her.

A few details of the proportions of a stoat help us to visualize the acrobatic performance that went on unseen behind the back of the settle. First, however, the name: the alternative is ermine, from Old French; stoat, a word of unknown origin, came into the language slightly later. One suspects that, as with so many medieval words, ermine was the name taken across the Atlantic by the Pilgrim Fathers, to become established in America and to fall into disuse in the country of its origin. In Britain the name ermine is reserved for the animal's white winter coat, ornamented with the black outer half of the tail, which does not turn white in winter, used on robes for royalty and the aristocracy. In North America the name is used for the living animal, and it is probable that in former times this was the more common usage in Britain, too.

The stoat, or ermine, then, is a near relative of the better known weasel, a member of the weasel family to which the ferret, polecat, marten, mink, wolverine, badger and otter also belong, as well as the skunk. All are short-legged and long-bodied, the long body being most emphasized in the smaller members of the family, such as the weasel and stoat. The stoat is a foot long with a tail less than half this length. The diameter of the body at its thickest point is about two inches, so Ermie's ability to manoeuvre so adroitly and speedily in the inch-space between the settle and the wall is little short of remarkable.

There is a saying that a weasel can pass through a wedding ring, though no doubt the alliteration is partly responsible for this. The short while we had a tame weasel Jane tried to test the truth of it. The length of a weasel varies a great deal from one part of its range to another but averages about eight inches, and the diameter of its body at its thickest point is well over an inch, whereas the diameter of a wedding ring is much below it.

Jane took a piece of board and drilled a hole in it the diameter of her wedding ring, but nothing would induce the weasel even to try passing through it. So she took the skull of a weasel, which she happened to have, and found this would pass easily through the ring. It is axiomatic that an animal will pass through any hole through which it can get its head — which merely serves to underline one of the constant problems facing those who keep animals in captivity. It is always surprising how small a space an animal can use to escape to freedom.

As soon as we had laboriously pulled the settle from the wall, Ermie would, as likely as not, dive behind a heavy seventeenth-century iron chest we have. That, in turn, had to be pulled away from its place against the wall. Having been deprived of two of her favourite hideouts she usually turned her attention to the curtains. These were heavy plush and extended from the floor to a pelmet just under the ceiling. All we would see would be small moving bulges in the curtains as Ermie ran up and down on the insides of them.

We dared not allow Ermie anything like total freedom or we should have risked disfavour among those of our neighbours who kept poultry. Since she was hand-tame we preferred to keep her in the house, although there were obvious drawbacks to having her loose in a furnished room! Moreover, we could only allow her that liberty under close supervision, for an animal so speedy and so adept at manoeuvring in so small a space might at any moment find an escape route we had overlooked. During her liberty periods for recreation nobody dared open a door, to enter or leave the room. The likelihood was that if a door was opened there would be a sandy-brown flash and Ermie would have been beyond recall — and sooner or later among a neighbour's poultry.

The reason why Nicholas had parted so readily with his pet was that his family lived in a flat and there came a time when his mother's patience was exhausted. This was something with which we could sympathize, especially when, on one occasion, Ermie took fright when she was being picked up from the floor and demonstrated her close relation to the polecat and skunk by squirting the contents of her anal glands onto the nearest wall in the sitting-room. Much soap was needed to cleanse the room of the unpleasant odour.

The first thing one comes to appreciate from having members of the weasel family under close observation is their apparent bone-lessness. I have had split-second flashes of Ermie with the front half of her body on its back with front legs in the air while the rear half of the body is in the normal position with the hindfeet planted firmly on the ground. This is more often seen in polecats in which the body is proportionately longer and the legs shorter. One pair of polecats, in play, would seize each other by the mouth or fore paws and roll over and over together, each corkscrewing the body in the process.

Occasionally one reads or hears of a stoat carrying off an egg. One such account described how a stoat was seen jumping down the steps of a henhouse, on its hindfeet and holding an egg clasped in its front legs against its chest. Naturally, we wanted to see what would happen if Ermie had access to an egg. A small pile of straw was arranged on the floor to represent a hen's nest and two eggs were placed in it. Ermie was placed on the ground beside the straw. She grew excited as she extended her nose, sniffing busily, towards the spot where the eggs lay hidden in the straw. She searched among the straw and finally found the eggs. Her first action then was to perform handstands over them, carried out, as were all her movements, at high speed. She pounced at an egg, putting her fore paws the other side of it, placed her nose under the nearside, then threw her hindquarters into the air.

Soon she tired of this and contented herself with scrabbling rapidly at an egg with both fore paws, as if she might be trying to scratch it open. She made no attempt to crack the shell with her teeth and it is even doubtful whether the scrabbling was not merely another phase in the play. The next phase included drag-ging one egg at a time backwards through the straw, holding it under the chin and between the fore paws. The precise way in which this was done was difficult to see, partly because the straw obscured the sight of it, but more importantly because of the speed with which it was done.

Then came the brilliant climax. Instead of dragging the egg

backwards she started to push it forwards with her nose. Even when pushing it through straw she travelled sufficiently quickly to make it difficult to photograph. When she had got beyond the straw onto the carpet her speed increased considerably. She piloted the egg with her nose with unbelievable speed right across the carpet. This performance was so spectacular that it seemed worth while persuading her to repeat it with a view to photographing it. When she had dribbled the egg right across the room it was picked up and placed in its former position among the straw. Robbed of her toy, Ermie ran back to the nest and again dribbled the egg swiftly across the room. This was repeated a dozen times and several pictures taken. All proved, when processed, to be out of focus. Finally, we decided to take a film shot: the egg was put back into the straw again but by now Ermie had lost interest.

How this incident can be related to the behaviour of the stoat in the wild is not easy to see. One thing that interested me more especially was that we had had enacted before us a piece of exceptional behaviour, exceptional at least to me. More strikingly, perhaps, it was not repeated. Several times subsequently eggs were made available to the stoat but she completely ignored them. It was the old business of tiring of a plaything.

The cage in which Ermie was lodged, at times other than those when she was let out for recreation, was kept in a room under my bedroom. Night after night I would hear downstairs the sound as of a wheel being turned rapidly. Investigation revealed that she was exercising herself in her cage, running round in circles. That is, starting from the floor of the cage she would place her fore paws on one wall of the cage, bring up her hindfeet, run across the ceiling of the cage and down the opposite wall. Always following this same route she would run round and round in this vertical circle for several minutes on end before coming to rest. The sound of her paws resembled the cranking of a wheel on its axle.

Years ago it was a common practice to keep a red squirrel in a cage fitted with a wheel, and the saying then was 'to run round and round like a squirrel in a cage'. It was no doubt something of the same kind of behaviour to that indulged in by Ermie that inspired the invention of the squirrel cage, with its wheel.

We once made an unintentional experiment regarding the value of a wheel to small mammals in captivity. Jane had acquired some African striped mice. She placed one pair in a large glass-fronted cage in her sitting-room. At least, there was a pair in it to start with. She had another pair in an identical cage in her studio. Her original idea was that having striped mice in the two rooms she used most she would have striped mice more or less continually within her field of vision, to observe their behaviour.

Jane's husband, Kim Taylor, is a zoologist, a specialist on the control of rats. He is also a natural handyman and between them they had fitted a wheel, a sort of treadmill, in the cage in the sitting-room. The motive, as Jane explained, was that they thought it would be fun to watch the mice should they decide to use the treadmill. (This, incidentally, opens up another line of speculation and observation on the question whether animals ever take decisions, although this is not the place to pursue it. We must be content for the moment with the assertion that there seems to be positive evidence that animals, like humans, do take decisions.) After preliminary hesitations and much inspection of the wheel, without actually making physical contact with it, both the male and the female ventured to place their feet on it. They found it would turn. Soon, either singly or both together, side by side, they were spinning it under their feet, the wheel turning merrily, the mice remaining stationary except for their rapidly moving feet.

Then Kim had the idea of turning his wife's whimsical impulses to scientific profit. He made a counter to be fixed to the wheel which would register the number of times the wheel was turned by the mice in a given time. He also contrived a second piece of apparatus which would record on a drum the same form of activity, thus giving a double check. The first counter was made to register up to 999 turns of the wheel, on the assumption, naïve as it turned out, that the mice would probably take a week to clock up this number. In the event, they turned the wheel 999 times in the first half-hour, so a second counter was made that would register up to 99,999. This figure was reached by the mice in just over a week. It was possible to calculate, therefore, that in the first ten weeks after the second counter was installed the mice had turned the wheel through a million revolutions. This was the equivalent of a pair of mice running 180 miles between them! Assuming they shared the wheel equally between them, this was the equivalent of 90 miles per mouse in ten weeks, or nine miles a week, which is just about the sort of distance mice are known, from other sources, to cover in their search for food, when they are free, in the wild.

The result was unexpected. The pair of mice tied psychologically to the treadmill produced litter after litter. The pair living in the studio under otherwise identical conditions showed no signs of breeding.

There is a link between this and the elephant in a zoo that began to suffer from digestive troubles, which puzzled those responsible for its care. It was noticed that the elephant, let out from its night quarters, went straight to a point in the fence bounding its exercise area and stood there all day waiting for visitors to give it food. The zoo authorities then banned feeding by the public so that the elephant no longer had any inducement to stand in one spot, with the result that it walked around its enclosure throughout the day. The digestive troubles disappeared.

From the beginning we followed the principle of giving a captive animal three things: adequate food, companionship and as much freedom of movement as possible. It is probably more than coincidence that visitors never fail to remark on how healthy our animals look. A small example of the third of these conditions, the maximum freedom of movement possible, relates to our foxes. Their pen was twenty feet by twelve feet. A fox running round the perimeter could cover only sixty-four feet before starting on a repeat run. By arranging branches and logs, the same animal had a second storey, so to speak, doubling the distance it could run without repetition of the substratum beneath its feet.

With the change in elevation comes a slight change of viewpoint, and so a variety of scenery. It is therefore possible to furnish a relatively small space so as to give it a sense of variety and freedom. The principle can be illustrated by reference to our present home, Weston House, a large and rambling building with two staircases, several landings and corridors, nooks and corners, small rooms and large. Any small child staying here, especially for the first time, derives entertainment merely from going up one staircase along the landings and corridors and down the other staircase, exploring the nooks and small rooms on the way. There is a feeling of adventure. And so it is with animals if the cages and aviaries in which they are incarcerated are suitably contrived and furnished.

There is another factor which I consider important. There is a tendency in zoos to give cages and enclosures a concrete floor because it can be easily cleaned. I have also known people who, like us, have had small private zoos, and have laid concrete floors in the aviaries and pens on the grounds that this is more hygienic. But we have always used bare earth, allowing herbage to grow on it, or covering it with straw, peat or sawdust, according to circumstances. We have made no effort to keep the floor clean, except to remove faeces when these are deposited in heaps, which some species do, or to remove straw or leaf-litter at intervals of several months. There is no indication that we have thereby allowed unhygienic conditions to develop and there has been no history of disease. Apart from anything else, natural ground is less wearing on the inmate's feet and the natural odours from soil and herbage must be pleasing especially to mammals.

The value of a near-natural substratum was demonstrated by Ermie. When she was first taken out, under strict supervision, after a fall of snow, she rolled in the snow, burrowed into it and somersaulted on it. Her actions gave the impression of a sensual enjoyment at having every part of the body in contact with the snow. It was as if the stoat enjoyed the feel of the cold, wet substance on her fur or, perhaps, on her skin. This was most strikingly brought out by one characteristic action, when Ermie dragged her underside along the surface of the snow, using the forelegs to pull the outstretched body while the hindlegs trailed behind. She seemed to be luxuriating in the new sensation, the feel of the snow seeming to send her into an orgy of twisting, turning and tumbling as if she had taken leave of her senses. She would also do the same on damp moss and other natural materials.

Our genet behaved similarly and fairly regularly, too, usually when, having been asleep all day, she came out for her nightly activities. She chose grass that was not wetted by dew or rain and there was less appearance of clowning, as compared with the stoat, because every action was slow and infinitely graceful. Speeded up, however, her performance would not have differed substantially from that of the stoat. Because it was carried out slowly there was more an appearance of the genet making a fuss of Mother Earth. In fact, she was making a fuss of herself on Mother Earth.

Domesticated and tame animals kept as pets are stroked and fussed or petted by their owners. One learns where best to stroke a particular animal to give it the greatest satisfaction. With a dog it is stroking under the chin or tickling behind the ears, with a horse scratching of the rump. So far as small mammals are concerned it is precisely those parts which Ermie and the genet endeavoured most to bring into contact with the snow, moss or grass, whichever was chosen for these luxuriating exercises.

We also had a mink for a while. This had a disused sink filled with water in its enclosure. After resting the mink would emerge from its sleeping quarters, enter the water and go through the most incredible 'aquabatics'. Jane tried to film this, unsuccessfully, largely because it was carried out at such high speed and with such a flurry of water. Watching the performance, however, after having seen Ermie and the genet luxuriating on land, one realized that the same movements were being carried out.

Otters, too, will at times roll on the surface in a similar way. The ensuing flurry of water effectively masks the fact that the animal one is watching is no more than a common or garden otter. Indeed, the monster of Loch Arkaig, recorded by Sir Herbert Maxwell towards the end of the nineteenth century, proved to be no more than an otter doing what we had many times seen Ermie, Jennie and the mink doing.

The Loch Arkaig incident concerned a huge animal which four gentlemen, friends of Maxwell, claimed to have seen rise from the depths of the Loch, create a tremendous commotion at the surface, then disappear beneath the water again. The Loch Ness monster had not yet hit the headlines and was virtually unknown outside the Highlands; in relating the story, Maxwell says that he had no alternative but to believe his friends' account of the monster. A few days later he happened to be talking to the stalker, a highlander who had accompanied the shooting party which included his four friends, and he asked him whether he had seen this mysterious animal. The highlander replied in the affirmative. 'What did you think it was?' asked Maxwell. 'It was an otter,' replied the stalker. 'Why did you not say so?' asked Maxwell. 'Nobody asked me,' replied the stalker.

The first photograph ever taken of what was alleged to be the Loch Ness monster, and which caused such an international sensation in the early 1930s, proves on closer examination, and especially when the picture is projected onto a screen, to be a common otter doing what we saw our tame mink doing many times in the water and Ermie and Jennie doing on land. Indeed, watching the antics of the stoat and the mink was one of the main factors that turned me from being an enthusiastic — even over-enthusiastic — believer in the reality of the Loch Ness monster to being a complete sceptic. We always meant to buy a large glass aquarium as a substitute for the porcelain sink in the mink's enclosure, to try for a film sequence of her spectacular antics in water. Unfortunately, when keeping a number of animals the expense is continuous and many desirable items of equipment have to be consigned to the future. Also, there just is not time for everything. So the large aquarium was not bought and in due time the mink was found dead in its sleeping-box, curled up as if asleep.

Another difficulty in filming the mink, or even taking still photographs, was its insatiable curiosity. He might be thoroughly enjoying his swim but if one went in with a camera, the likelihood was that he would cease operations in the water until he had fully inspected the camera and the person holding it. Even then, after he had returned to the water, any movement by the photographer might arouse his curiosity further and he would stop again to see what else was afoot. Ermie showed a similar sense of curiosity but to a lesser degree.

The old-time naturalists used to speak of certain animals using what they called 'charming'. A fox, for example, might be seen prancing on stiff legs, rolling on the ground or somersaulting, as if it had gone crazy. Rabbits and birds would draw near to watch, drawn by curiosity, until they formed a ring round the fox. Finally the fox would pounce, scattering his audience except for the one he held in his mouth or between his paws. Similar scenes have been described for stoats and martens but, in my opinion now, the central figures in such episodes were doing no more than we have seen in our stoat, genet and mink — luxuriating in the feel of herbage on their fur or skin. The rest of the explanation involves the sense of curiosity of those animals that gathered around to watch this unusual exhibition.

Curiosity is a compelling attribute of the higher animals, and it goes further than merely putting the subject *en rapport* with its surroundings. Under test conditions, monkeys and apes have been seen to leave a favourite food in order to satisfy their curiosity. They will suspend the meal entirely in order to go to the opposite side of the cage to peer uncomfortably through a small window, if there is any sound of movement outside. It is very like the habit of certain people one hears about who cannot resist peeping through the curtains to see what their neighbours are doing.

There is an obvious biological value to wild animals in having a developed sense of curiosity except where, as with the pronghorn of North America, it can elude the hunter by its speed but can be enticed back within gunshot range by its own curiosity. The hunter simply waves a white flag, or does something else unusual and the pronghorn comes back to investigate. In ourselves this deeply rooted impulse has led to a variety of accomplishments and is the basis of what we call the spirit of inquiry; and in most members of the weasel family, too, it seems to be far stronger than is necessary merely for purposes of successful living. It has become almost an obsession with them to see what is going on and many a stoat or marten must have lost its life through its inordinate desire to see what the man with the gun is about to do. I once sat for half an hour facing a wild stoat. I wanted to see what the stoat would do. The stoat was equally busy watching me.

It was while these and other thoughts were filling my mind one day that my wife drew my attention to something that was happening in the garden, immediately outside my study windows. So I was able to watch the whole panorama without risk of disturbing the *dramatis personae*. In a cage twelve feet by six feet by ten feet high we kept a pair of grey squirrels. One was on the ground playing in the same manner as Ermie and the others, that is, it was going through the evolutions which, in fox, stoat and marten, were labelled 'charming' by the older naturalists. A bantam cockerel, free in the garden, had wandered over to watch. He stood with his beak against the wire-netting of the squirrels' cage, seemingly spellbound. In a very short time he was joined by other birds, some of which had come over quite purposively from as much as fifty yards away, in a straight line to the cage. In the end there were, in addition to the cockerel, two bantam hens, nineteen half-grown bantams, a large white leghorn, three black East Indian ducks, two mallard and two Muscovy ducks. They all stood in a straight line along the front of the cage, beaks to the wire, and for ten minutes watched intently as the squirrel played.

Substitute stoat for squirrel and wild birds for domesticated birds and the scene was typical of a so-called charming. This would then give us a complete explanation for this otherwise puzzling phenomenon, even to the fact that some of the squirrel's audience looked a little dazed, almost mildly intoxicated, which is one of the features mentioned at times by writers on the subject. That the stoat, marten or fox goes on to seize one of its audience can then be interpreted as sheer opportunism.

When Nicholas brought Ermie to her new home he had demonstrated for us how he could hypnotize the stoat. He held her by the neck with the thumb and index finger of his left hand, letting the body hang freely in the air. Then he stroked her gently with the right index finger from the tip of the nose to the forehead three or four times. After this the stoat hung limp, as if in a deep sleep, and could be swung gently in circles, giving no response whatever. To have swung her in this way normally would have brought a vigorous scrabbling with the legs and a struggle to escape from his grip.

A few days later we had a naturalist visitor. We showed him the stoat and, among other things, told him about the 'hypnosis'. Later we took him to see our other animals. When we came to the owls' aviary he asked whether we had ever hypnotized an owl. We had not – so he went into the aviary and gently lifted one of the owls off its perch. The first thing it did was to drive its talons through his shirt-front and into the skin of his torso. Having gently disengaged the talons, wincing as he did so, he held the bird under one arm, supported its head with the hand of that same arm and, with the thumb of his free hand, stroked the upper surface of the owl's beak gently from the tip to its base, three or four times. The owl, to all appearances, was fast asleep by then. Our visitor laid the owl on its back on the ground in a corner of the aviary where it lay as if dead for three minutes or so, after which it turned onto its feet, shook itself and flew back onto its perch. It was none the worse for its adventure.

Both these are variants of the long-known trick of holding a chicken's head to the ground and running the finger in the dust in a straight line away from the beak. After this the chicken remains quite quiet, as if in a deep sleep, for a while before coming to, shaking itself and resuming normal activities as if nothing untoward had happened. The first recorded experiment in this field was carried out in 1646 by Father Athanasius Kirchner, who took a partridge, bound its legs together, laid it on the ground and drew a chalk mark from the tip of its beak along its body. When he removed his hand the partridge lay as if dead. Kirchner called this his *Experimentum Mirabile*.

It has since been found that merely turning an animal onto its back quickly will produce the same effect and that this state of akinesis (literally, without movement) is very similar to the group of phenomena which include shamming dead or thanatosis, when an animal drops to the ground and remains immobile, to all appearances dead. However, it is one thing to give technical names and another to understand fully what is happening and this animal hypnosis, akinesis or thanatosis is even now but imperfectly understood.

We had only two comparable instances among the many animals that passed through our hands. One was in a bullfinch. Veronica Watts, daughter of a neighbour, volunteered on one occasion to feed our animals when we were on holiday. One day, as she entered the aviary containing a pair of bullfinches, the cock flew into the space between the electric light bulb and its shade, the light being there to illuminate the aviary. It became wedged and Veronica gently extricated it. The bird lay on its back in her hand, limp and to all appearances dead. While she was wondering what she would have to tell us when we returned and found she had failed in her responsibilities, the bullfinch stirred, seemingly came to life again, got onto its feet and flew onto a perch, none the worse for the misadventure.

The other incident concerned our bantams. Two of the hens were sharing a nest, sitting side by side, brooding twenty-four eggs. One morning, Bernard Fry, our second gardener, went down the garden at nine in the morning. He saw three chicks in the run, so the eggs had started to hatch. Unfortunately, two of the chicks lay dead with lacerations on head and body. The third was lying apparently dead, limp and cold to the touch, but with no visible injury. Bernard threw all three onto the compost heap.

It was evidently the work of rats since there was a rat hole near the nest and only sixteen eggs remained in the nest. Perhaps the rats had been disturbed by the arrival of Bernard and our head gardener Wally Fry, who now set to work pulling weeds from around the compost heap. Three-and-a-quarter hours later, one of them gathered up an armful of weeds, threw them on the compost heap and heard the cheeping of a chick. He thrust his hands into the weeds to part them and disclosed the 'dead' chick, the one without injuries, alive, vigorous and calling. This was the first they had heard of it although they were working close to it. The chick grew into a handsome adult bantam after over three hours akinesis. It was subsequently named Lazarus!

7 Niger, modern phoenix

In the early summer of 1956, Jane and I went to call on Ivor and Audrey Noël-Hume, friends who lived in the upper half of a two-storey house in Wimbledon. They kept tortoises, a dozen or more of them, which had the freedom of the flat by day but were let loose in their garden for exercise and sunning on summer evenings. Each tortoise had its own little wooden hutch in the garden in which to sleep at night and through the winter.

I had vivid recollections of our previous visit. After dinner the four of us had sat in front of the fire talking and in front of us the tortoises, of various species and sizes, ranged themselves in a semi-circle enjoying the warmth. Had I had a crystal ball then I should have seen that this fire-worshipping by the tortoises almost qualified for 'coming events casting their shadows before them'.

On this latest visit we found Audrey in a somewhat disturbed state. 'I'm expecting Ivor home soon,' she almost blurted out, 'and I don't know what he is going to say. The other day he said to me firmly "No more pets". Today I took a fancy to a fledgling rook that somebody was hand-rearing and I've brought it home with me.'

Having said this she produced the rook and put it down on the carpet where it strutted about confidently, as if it owned the place. It came first to one and then the other of us, and at each person it paused, opened its beak and solicited food. The fact that neither of us offered it any seemed not to deter it. Indeed, its gaping seemed almost an act of friendliness.

At that moment we heard the street door open and shut. Ivor was home. We had no time to wonder whether we were about to witness a domestic argument. He came into the room, took one look at the rook, which strutted over to him and opened wide its beak, and sat on his hunkers to take a closer look. To the best of my recollection Ivor took little notice of us – he may have said a cursory 'Good evening' or he may have briefly nodded in our direction – but I can clearly recall his face at that moment. It was radiant. The rook stayed.

For the rest of our visit that evening the rook was the centre of attention.

If Audrey had erred in bringing the rook home she paid the penalty, for there could surely never have been a bird that was more humanly-fixated. Audrey built an aviary for it in the garden but Niger, as she had named the rook, refused to do without human company. Whenever she left it to go indoors, the bird called so vociferously that she was afraid her neighbours might complain. She was compelled to take a chair out into the garden each day and sit beside the aviary, talking to Niger to keep him quiet. She spent almost the whole of each day in this way, whatever the weather, and she had to sit there each evening until the light faded and Niger had settled down for the night.

Audrey told me all this when she brought the rook to us to hand him over to our care, which was only a few weeks after that memorable evening when Ivor and Niger had first faced each other on the sitting-room carpet. It must have been almost providential that Ivor should have been invited to take up an appointment in the United States. At least it freed Audrey from the bird's tyranny. This turn of events was equally providential for me, for it set me firmly on the trail of the phoenix, and led to much exciting research to find an explanation for a myth that had baffled scholars for centuries.

There must be few people in the western world who are not familiar with the name of this mythical bird or of the saying 'To rise phoenix-like from its ashes'. The picture, too, of the head and shoulders of a bird rising above flames that have engulfed its body must be hardly less familiar. The full story has, however, been obscured by time and this alone took much effort to search out.

Briefly, the story of the phoenix dates back to Herodotus, the Greek scholar, in the fourth century BC, who said that this rare bird appeared from the east once every five hundred years. It built its nest in a palm tree in the Levant, set it on fire and was consumed to ashes. Three days later, a small worm was seen among the ashes, and this grew into a new phoenix. The resurrected bird then gathered 'the ashes of his father' into an egg-shaped mass, wrapped this in myrrh and enveloped the whole in aromatic leaves. Then it flew with the egg-shaped mass in its beak to the temple of Heliopolis, in Egypt, laid it on the altar, bowed twelve times to the sun, and flew away to the east, only to be seen again in another five hundred years.

I do not remember clearly now what started me on my search. I had read several things in books, which began to crystallize when Richard, then doing his national service, sent me a cutting from a newspaper, describing a fire in a building in Harrogate. It was believed to have been started by a bird taking a lighted cigarette to its nest on the rafters and setting it on fire. This struck me as odd and when, a few days later, the news editor of a national newspaper rang me I asked him whether he had ever heard of such a thing. He replied that when he was a reporter 'we took this for granted'. So here was something apparently well known to pressmen but completely unknown to the world of zoology.

Of the many astonishing things that Niger taught me, this was perhaps the most astonishing: that in the voluminous literature on birds that has been published this century there has been no mention of this fire-raising habit. I remember meeting one of our outstanding ornithologists (at a commemorative dinner in London), at about the time we made the acquaintance of the Noël-Humes. He had described ornithology as a sucked orange. When I asked why, his reply was: 'Everything about birds has been studied in detail: how they eat, how they breed, how many times a day they preen, defaecate and so on.' He had overstated the case, although it was very near the truth, yet here was an important aspect of bird behaviour that, so far as I could see, had been entirely overlooked. And there was no excuse for it.

Pliny tells us that birds (they were probably jackdaws, but of this we cannot be certain) carried burning embers onto the thatched roofs in ancient Rome, setting the houses on fire. He also mentions, in another place in his *Historia Animalium*, that a false-phoenix had been seen in Rome. There is no more to it than this, but presumably he was referring to a bird in some way playing with fire. In the Great Chronicle of London, for 1203, it is recorded that birds were seen carrying glowing embers onto thatched roofs, so spreading the flames in one of the many fires that have devastated that city. In books of the sixteenth and seventeenth centuries various members of the crow family are branded as *Aves incendiaria* (firebirds) because they set thatched roofs on fire. There are also several pictures in books, from the sixteenth century onwards, showing birds enveloped in flames. They are supposed to be illustrations of the phoenix but their interest lies in the fact that they are not typical of the early accounts of the 'mythical' bird.

So there was a fair amount of documentary evidence suggesting an association between birds and fire. More was to follow and, as

often happens when one is on the trail of a problem, accident
took a hand. A neighbour had invited me to call on him to see
the wild chaffinches he had encouraged to come into his house
and take food from the table. Jane and Richard were with me.
Then I noticed a double row of plaques, formerly used by fire
insurance companies, around the walls of the sitting-room, including four used by the Phoenix Insurance Company.

I asked my host if he was interested in insurance or in fires. He
replied that he had been an officer in the Guildford Fire Brigade.
So I put the question then uppermost in my mind: 'Have you ever
heard of a fire being caused by a bird taking a lighted cigarette to
its nest, setting the nest on fire and thereby setting the house on
fire?'

'Of course,' was the immediate reply and I was given instances
from his experience, including descriptions of trees ablaze in their
upper part, standing in the middle of fields well away from any
possible sources of fire.

Later I was to find that such events are well known to fire ser-
vices in various parts of the world and, to digress for a moment, I
ultimately came upon tangible evidence of this when we moved to
Albury. Soon after we took up residence there, Mr R. J. Howes,
then head forester to the Albury Estate, took me on a tour of his
woodlands. We fell to talking about the origins of forest fires.
Inevitably I put to him the question I had asked so often before:
'Have you ever heard of a bird taking a lighted cigarette or other
burning materials to its nest, and setting it on fire?' Mr Howes
then told me the following story.

One of his woodmen was walking with his wife one Sunday
afternoon through Weston Woods, which is near my present
house. At one point along their path they passed a yew tree. The
woodman, taller than his wife, had to duck his head to pass under
a branch that hung over the path. As he did so he felt a burning
sensation on the back of his neck. He looked up to see flames and
wisps of smoke in the heart of the tree. By the time the firemen
arrived the upper half of the yew tree was fully ablaze.

At my request, Mr Howes took me to see the yew. The upper
branches were charred, blackened and bare of leaves and at the
base, in a crotch formed where these branches left the trunk, were
the charred remains of a jackdaw's nest. The nearest house, a
lonely woodman's cottage, was over two hundred yards away.
Something must have transported burning materials to the nest.
In this lonely part of the woods where only woodmen went, it

could only have been the jackdaw.

On our way home after our visit to the ex-fire officer's chaffinches, Jane, who throughout the conversation about birds and fire had maintained a total silence, broke the news of a film she had shot. She was then a young woman of nineteen, unassuming and somewhat shy, and she told her story with diffidence and hesitation, but with an evident air of suppressed excitement. She must have been bottling it up all the time I was talking to our neighbour. She had seen me, she explained, testing Corbie (see Chapter 5) with burning straw so, unbeknown to me, she had placed a handful of straw in Niger's aviary, put a match to it and filmed what he did. She had then sent the film for processing and she was expecting it back that day.

As soon as we were back home, Jane took more straw into Niger's aviary and set fire to it. As we watched I realized we were on to something big. It would be hard to say which of us, Jane, Richard or myself, was the most thrilled, for none of us showed much emotion while watching. I can speak only for myself – I was almost beside myself with excitement! The moment Jane went into his aviary with a handful of straw, Niger became wildly excited, and being only a bird made no attempt to hide it. He hopped around Jane's feet and stretched his head up to her expectantly. Jane stooped and placed the straw on the ground. She struck a match to set it on fire. The rook snatched the burning match from her hand, raised his wings and held the flame under his left wing. She struck another match and this he also snatched from her and held under his wing. Eventually she managed to set the straw alight.

Niger snatched at the burning straw, spread his wings and twisted his tail to one side in what, by now, we could readily recognize as the typical anting posture. He jumped on the straw, all the time with his wings brought forward and his head constantly moving, snatching at the flames and smoke and passing his beak first under one wing, then under the other. There is no cruelty involved here. No compulsion was used. And although Niger stood on the burning straw longer than one would have thought possible without injury, there was no smell of charred feathers nor signs of singeing. Indeed, throughout the many scores of times Niger 'played' with fire he never sustained injury, and he lived to a ripe old age.

While watching this extraordinary display the idea was born that his behaviour might hold the solution to the age-old myth of the phoenix. Jane's film arrived back from processing that day so we were able to project it and study carefully a performance such as we had witnessed that morning. After seeing the film I suggested that we ought to persuade Niger to repeat his actions above eye-level, to simulate a bird in a tree. We might then have the sort of scene that gave rise to the traditional picture of the phoenix.

We drove a post into the ground and fixed a horizontal board to it five feet above ground level. On this we placed some straw. Then the ciné-camera was set in position, Niger was brought to the site and the straw ignited. Immediately, he flew up to the straw and, seemingly surrounded by flames and smoke, he spread his wings and as he turned his head to one side to put his beak under his wing, we had before our eyes the centuries-old, traditional picture of the phoenix, in every detail. The essential part of the phoenix story, which dates back at least two thousand years, to Ancient Egypt, was being re-enacted before our eyes. The phoenix was reborn.*

As can be imagined, in the early days of my researches into the phoenix story I had put the question 'Have you ever heard of a bird carrying burning materials to its nest?' again and again to one person after another who might conceivably have further information to offer. At a very early opportunity I put it to one of our outstanding authorities on bird behaviour. He is normally gentle and mild-mannered but on this occasion his reaction was immediate, almost violent. For some reason, unknown to me, he became quite heated, declaring that it was quite impossible that a bird should do this and that there was nothing in the known principles of bird behaviour which could make such an action possible. As he warmed to his harangue, he gave me a detailed analysis of bird behaviour during nesting time, together with any number of reasoned arguments, to emphasize why no bird would ever carry a lighted cigarette or a lighted match to its nest.

I tried other students of bird behaviour with similar results, so it is safe to say that no student of bird behaviour nor animal psychologist would, at that time, have regarded the testimony of a

---

* Much of what has been set down here, together with other aspects of the story, has already been set forth in my book, *Phoenix re-born*, published in 1958. It has been necessary to recount once again some of these details in order to add some of the more personal details that were omitted from the book, as well as to provide a secure base from which to stress some of the additional parts of the story of Niger. Above all, I would wish to emphasize the compulsive or obsessive features of Niger's behaviour towards smoke and flames.

fire officer as anything but nonsense. And this also explains why everything associated with the phoenix story, or linking birds with fire, was regarded as pure unadulterated myth or legend. So an important principle emerges: that it is unsafe to eschew sober and serious claims by witnesses of animal behaviour solely because it cannot be brought within the orbit of the known principles. It may be that all the principles of animal behaviour have not yet been set forth.

It may be not without point to mention an experience I had a year or two before Niger flew into my ken. I was walking with my wife through the countryside when we came across a grass fire sweeping across rough pastureland. We stood watching this with the same kind of fascination that apparently draws Niger towards burning straw! The line of flames reached up to six feet in the air, so it was not a devastating fire although slightly terrifying to watch at close quarters. As we stood there, the flames engulfed some scrub at one point and it was then I noticed a bird, either a thrush or a blackbird, persisting in flying over the flames, partially hovering, wheeling to fly away, then coming back. It repeated this many times. I recall saying to my wife: 'I guess that bird has a nest in there somewhere and is trying to get back to it in an attempt to save its eggs and nestlings. Doesn't it make your heart bleed?' This may have been the correct interpretation, but I am more inclined now to the view that we were not witnessing a display of heroism but a bird anting with fire.

Niger continued to be a showpiece for visitors to our menagerie and when we moved to Weston House he was given a larger aviary. Oddly, although he had never escaped from his aviary at Horsley he did so a number of times at Albury. The first intimation that he was out usually came when somebody telephoned. It became a familiar message: 'There is a large black bird sitting on my window sill tapping the window pane as if asking to be let in. Have you lost one of your birds?' A visit would reveal that it was Niger. He was so human-fixated that as soon as he had regained his liberty he would compulsively seek the company of humans. Moreover, it was no trouble to pick him up and bring him home.

Then a housing estate was built less than half a mile from where we lived. Up to that time, Albury had been mainly a village of elderly people. The newcomers were much younger, and in the crescent-shaped street that bisected the estate a dozen or more children could be seen playing outside school hours. Now, when Niger escaped, he always made straight for the crescent, and we began to receive telephone calls from one of the mothers: 'There is a large black bird playing with the children. Have you lost one of your birds?'

I would arrive to see Niger strutting about at the centre of a ring of laughing children, and it now became less easy to recapture him, largely it seemed because he enjoyed being with children. There was one occasion, for example, when he stayed where he was until my hands were about to close on him. Then he flew up and away, to land a short distance from me. After he had treated me to these evasive tactics for a while, I went down on all fours and crept up to him, recapturing him without difficulty. This is an old trick — animals are less afraid of somebody coming down to their level — but it was a trick I would have preferred not to use before an audience of children. Instead of the derisive laughter I had expected, however, the children seemed to be looking at me with greater respect, because they had tried to recapture Niger and failed, whereas I had succeeded.

There were times when Niger would escape from his aviary and come back of his own accord. On such occasions he would reappear as evening drew on and land on the top of his aviary. No doubt he would, sooner or later, have gone back into the aviary to roost, but I felt it unwise to risk leaving him there. I tried to lure him into the aviary with food. This was unsuccessful. Then I tried to capture him by enticing him within reach of my hands. This was frustrating because Niger would come as far as my finger-tips and then move away as I advanced my hands to take hold of him. It would not have been difficult to recapture him by a sudden grab, but this would have been to risk giving him a fright and so reducing his tameness.

Then I had a brainwave. I fetched some straw, went into the aviary, set fire to the straw and came out again. I nearly collided with Niger in the doorway, he was in such a hurry to reach the straw. On another occasion when he had escaped he sat near the top of a tree forty feet up from the ground. We called to him, he replied *wha-ha* to each of our calls but he made no attempt to descend. Burning straw in his aviary soon brought him down.

Even at normal times, while Niger was in his aviary, he would grow excited and rush over to the wire-netting when one carried no more than a wisp of straw past him. His wings would be half-spread and his tail partially twisted to one side in a half-anting posture. He would do the same if I merely took a box of matches from my pocket within his sight, just as Corbie did.

Perhaps we should have been more adventurous with Niger and allowed him his liberty. We were always afraid that, given his freedom, he might pester people, possibly scare them, and then be shot as a nuisance. But there was also a free-living rook that helped us fill in the other side of the picture. This was Reggie. We made his acquaintance through Mrs C. Pearce, then living in Sussex. Her sons had found Reggie as a bedraggled fledgling that had fallen from the nest. They took him home and hand-fed him, but at no time was he kept captive in the usual sense. When able to fly he joined the rooks in the nearby fields but returned to the Pearces' cottage for breakfast each morning between 8 and 9 a.m. He also showed a fondness for snatching lighted cigarettes. He would then spread his wings and hold the cigarette first under one wing, then under the other.

We made several early morning visits to see him but on each occasion he failed to turn up as usual. On our last visit he still had not appeared at nine o'clock. We waited and waited for another hour. Then, just as we had decided to depart, in the hope of returning another day, Reggie flew in. Once inside the house Reggie went straight to the plate of food put for him on the floor and was soon busy eating. Having satisfied his own appetite he started to fill his throat pouch, which, according to Mrs Pearce, he had been doing each morning since the courting season began, in order to go back and feed his mate.

Here, then, was our dilemma. We had already had the novel experience of seeing a tame rook that voluntarily visited the abode of his benefactors. That should have been enough to satisfy us and we ought to have been willing to let Reggie depart on his lawful and quite natural mission. Yet we also wanted to observe the actions of another rook addicted to burning materials.

I lit a cigarette and held it towards Reggie. He looked at it and was clearly interested but he made no attempt to take it, for the simple reason that he was unable to do so, and he knew it. His beak as well as his throat pouch were full of food. He looked at the cigarette for a few seconds, as if trying to make up his mind how he could have the best of both worlds. Then he turned his head away, pointed his beak downwards and let drop some of the accumulated food. He picked this up piece by piece and swallowed it in the normal way. It was as if he appreciated that it was impossible to fill his beak with food and at the same time indulge a favourite 'pastime'.

We waited patiently for the end of this stage in the proceedings before offering him another lighted cigarette. Once again he showed mild interest in the cigarette but made no attempt to take it into his beak. On the contrary, he found some more food and again filled his throat pouch and beak. There could be little doubt that whatever fascination a cigarette may have held for him he had only one impulse now: it was to take food to his mate.

Finally, after several repetitions of the sequence, of emptying the beak and eating the food, as if clearing the way for performing, he took the cigarette. Even so, he showed little enthusiasm for this pursuit and dropped one cigarette after another on the floor without performing. In the end, Mrs Pearce took one of the lighted cigarettes saying 'Come now, Reggie, I wonder if you will perform if I put it under your wing for you.' She did precisely this. The rook immediately spread his wings, in the typical anting manner, but did not place his head under the wing. On the contrary, he held it as if about to rub the back of it on his shoulders.

Having thus performed once, stimulated by the hot end of the cigarette near to but not touching the plumage, he continued to give us six more separate performances. In each of these he himself took the cigarette and held it under a wing. The seventh time he was offered a cigarette he took it in his beak and dropped it on the floor. The impulse to ant had died down although he continued to peck at the smoke rising from a proffered cigarette. Finally, he once again filled his throat pouch and his beak with food and showed by his every movement that his one aim was to fly away to his mate, which he eventually did.

Some people might consider, and not wholly without justification, that our treatment of Reggie came into the 'dirty tricks' category. At least the results seemed to confirm the compulsiveness of anting in general, and anting with burning materials in particular. They also suggested that the important stimulus lies not in the materials held in the mouth but in the heat applied to the feathers. This we had suspected in Niger's performances, and with Corbie, but there had never been so clear an indication as we had from Reggie.

The question has often been put to us: 'How is it the bird does not hurt itself?' This is a natural question because a bird, like Niger, anting with fire seems to be in constant peril during this spectacular performance. There are, however, several safeguards. First, the white nictitating membrane, the so-called third eyelid, is drawn across the eyes all the time. Secondly, at the onset of anting the mouth is seen to be filled with saliva. Thirdly, although one speaks of the bird spreading its wings there is also a flapping movement which is probably just sufficient to keep the tips of the feathers away from the flames.

This flapping probably also explains how a bird, having taken an ember or a lighted cigarette to its nest, manages to set it on fire. One year, when the breeding season was at an end, I collected a score or so of disused birds' nests, made sure they were bone-dry and on a hot day, to give optimum inflammability, spent some time putting smouldering cigarettes in each nest. None caught fire. At most there was slight charring of the nesting materials adjacent to the business end of the cigarette. Eventually I tried a gentle fanning with my hands to simulate the flapping actions of a bird's wings while anting. Only then was actual flame produced.

I shall never forget the evening when I communicated the results of my researches into the phoenix story to the assembled Scientific Fellows of the Zoological Society of London. When the Chairman called upon me to deliver my discourse, I walked up to the dais, turned to face my audience, squared my shoulders and flexed my arms so that my forearms rested approximately where my diaphragm would be, and locked the fingers of my right hand with the fingers of my left. It is a posture I have always associated with a prima donna singing and I have always assumed this posture results in keeping the rib cage free for the deep breathing necessary to produce the full volume of voice. This may have looked like a theatrical pose to members of my audience but I was doing it deliberately. I knew that I would be describing Niger spreading his wings and flapping them before the flames. There would be the temptation to gesticulate in order to emphasize my description of the bird anting with flames. This would have led me to wave my arms about like a maniac. I was determined to avoid this even if it meant posing like a prima donna!

The following day, my colleague Peter Crowcroft, now Director of the Chicago Zoo, met me in the corridor in the Natural History Museum and said: 'I very much enjoyed your talk yesterday evening on the phoenix. When I saw you stand up confidently and fold your arms across your middle I knew we were in for something good.'

He was not to know I was only trying to stop myself looking like a flapping phoenix!

Just before midday one spring morning a small figure was seen bustling through the front gate and towards the house. It was little Mrs Fulbrook, and the soubriquet 'little' is well merited, for she could not have been more than four feet tall. She was carrying a large cardboard carton. Moreover, she had walked the best part of a mile with it, such was her devotion to dumb animals, for inside the carton was hay and inside that a female hedgehog and her two babies.

It seems that the farm machine Mrs Fulbrook's son had been working had inadvertently destroyed the nest in which this small family was sleeping. Fortunately, the animals themselves were not injured but, fearing that if nothing were done the mother hedgehog might desert her babies, Fulbrook junior gathered them up, took them home and, in conjunction with his mother, decided we were the best people to look after them.

I have always been mindful of Mrs Fulbrook's humanitarian act, and grateful, too, for as we shall see it was a most fortunate event for me.

If we say a hedgehog is as large as a person's head we shall not be far wrong. Heads vary in size and so do hedgehogs. Even the individual hedgehog seems to vary greatly in size depending on the time of year and on what it is doing. When walking about unruffled, with its coat of spines lying flat and sleek on its body, it looks relatively small. As soon as it is disturbed it rolls into a ball and its spines bristle so that it looks considerably larger.

The hedgehog is one of the *Insectivora*, an order of mammals that also includes shrews and moles. As an order the *Insectivora* are among the most primitive of mammals. Indeed, the earliest known fossil mammals are shrews, but although hedgehogs are related to them they must have put in an appearance much later in geological time. Shrews are found today throughout Eurasia and Africa, and also in America, but hedgehogs came into being after the Americas had separated from the Old World land mass, for they are not found in the New World.

A number of mammals have a spiny coat – porcupines, for instance, and also tenrecs and spiny mice – but it is doubtful whether any have found so secure a place in the affections of the human race as hedgehogs have, and this goes especially for the European hedgehog. At all events, in Britain there is probably no other mammal that is pampered by so many people, even though not many are kept as pets.

A hedgehog is harmless and inoffensive. It does no harm to crops and it does not bite if handled. It merely rolls into a ball. Far from damaging plant crops it feeds largely on pests like slugs, and on earthworms. It also has the advantage of looking quaint, almost antediluvian, and there must be tens of thousands of people who regularly put out a nightly bowl of bread and milk for a favourite hedgehog during spring and summer. (There is no need to do so in winter since hedgehogs hibernate.) The animals appreciate these favours for there are many reports of hedgehogs, newly emerged from hibernation, coming to the same back door as last year and waiting for their bowl of bread and milk. Some are reported to scratch at the woodwork of the door as if asking to be fed.

Throughout history, from at least the time of the Ancient Greeks, hedgehogs have figured prominently in European folklore. They have also provided food for gipsies, who are said to pack them in clay and put the whole into a red hot fire to roast; when the clay is removed the spines come away too. It was not long after pheasant preserves became fashionable that hedgehogs were accused of eating the eggs from under sitting hen pheasants and were accordingly shot in large numbers: parish records for centuries past are full of entries about the numbers of hedgehogs killed and of the bounties paid for their slaughter. With the coming of the speedy automobile came another devastating hazard – death on the roads. Hedgehogs can run quite fast when they choose to do so, at least as fast as a rat, but normally they choose not to, and do not use speed to escape danger. Instead they freeze, coming to a halt and remaining stock still with spines raised. When danger is imminent they quickly roll into a spiky ball, but this is useless against the wheels of automobiles. Thousands die each year through trying to cross main highways at night, for the hedgehog is nocturnal – it comes out at twilight, feeds throughout the night and returns to its sleeping quarters at dawn.

All this goes to show that in their native countries hedgehogs are well-known and familiar animals, living on our doorsteps. Yet until a few years ago little was known about them except legends, until Konrad Herter wrote his voluminous monograph on them. Which brings me back to Mrs Fulbrook, for in the year in which she arrived with her gigantic carton I had been on holiday in Cornwall and anticipating a leisurely time when I planned to do little more than sit on high cliffs looking out to sea, I took Herter's book with me. I can read German, with difficulty, and this was my opportunity to work my way through his hefty tome. And it was in this fascinating volume that I first learned of the hedgehog's habit of *selbsbespuchen*, literally spitting on one's self.

In this, a hedgehog is seen to select an object and start to lick it; and as it licks its mouth fills with foam. Then it turns its head, throws its body into contortions and, putting out a long tongue, deposits the foam on its spines. After this it resumes licking until the mouth is again filled with foam and again it plasters the foam on another spot on its spines, its tongue coming out a half a dozen times for each spot in turn. It may go on plastering itself with spittle for as much as twenty minutes at a stretch until its coat of spines is a dripping bedraggled mess of foamy saliva. 'Spitting on one's self' is expressive and descriptive, but cumbersome. I wanted a simple name for it in English and coined the word 'self-anointing', which has become generally adopted.

To me the amazing thing was that there was not one word about this remarkable piece of animal behaviour in the whole of the English literature, apart from one or two fleeting references. Indeed, so far as I can discover, it was quite unknown anywhere in the world until 1913, when a brief note was published in a German scientific journal; and the only other mention up to 1956, when I wrote about it in the *Illustrated London News*, was in Herter's monograph published in 1938. Yet the hedgehog is, as I have already emphasized, a widespread and familiar animal in the Old World. The gipsies knew about the habit, as I later discovered, and so did a number of ordinary people. For example, somebody once wrote to me saying her pet hedgehog had the habit of licking the legs of the sideboard and then going into convulsions, foaming at the mouth. This was before I read Herter's account of it.

Naturally, I was all agog to see this remarkable process for myself; and Mrs Fulbrook's hedgehogs showed it me. Jane, at that time, was busy with a ciné-camera, filming the animals we were accumulating. She wanted to film the mother hedgehog with its babies. The babies were tame enough but the mother simply rolled herself into a ball of prickles and refused to co-operate.

Moreover, Jane found it impossible to film the babies because they just would not stay in one spot; they were constantly moving out of focus.

'Can you spare a moment?' Jane called to me.

'What is it?' I replied.

'Every time I try to film this baby hedgehog it wanders quickly across the table, out of the range of the camera. I want you to sit with your hand on this table, with a baby hedgehog in it. Hold it in your open palm, to keep it in range of the camera.'

Hardly had the baby hedgehog been placed on my hand when the tiny mite started to lick the tip of one of my fingers. Joy, oh joy! It began to foam at the mouth and then, slowly and deliberately, it turned its head towards its back and proceeded to place the foam on its soft, infant spines.

There are few thrills to equal the thrill of discovery, even when it is a trivial one. At that moment I realized I was witnessing something on which the whole of the English literature was silent. I strained my eyes so as not to miss even one tiny detail of what was going on. I forget whether my hand actually trembled but inwardly I felt 'like stout Cortez when with eagle eyes he stared (for the first time) at the Pacific – and all his men looked at each other in a wild surmise, silent upon a peak in Darien'.

'Did you manage to film it?' I asked, fearful that the action might not be repeated or that Jane might, like me, have been too intent on watching it and so failed to manipulate the camera correctly.

'It's all right,' she reassured me, 'I'm fairly certain I got the whole of it.'

I need not have worried. We had dozens of hedgehogs in our care in the years that followed and there were very few that failed to give us an exhibition, at some time or other, of self-anointing. By far the most remarkable 'self-anointer' I ever met was Rufus, the pet of Mrs Castleman, of Northampton.

Soon after my notes about the baby hedgehog appeared in print I received a letter about Rufus, inviting me to go and see him. Instead, Jane went on her own and filmed Rufus. I was so impressed, when this film was projected, I decided on a second visit. Jane came with me and filmed another sequence. For a full three hours that Saturday afternoon we followed Rufus about Mr and Mrs Castleman's house, with camera and floodlights. First, he licked a spot on the carpet and self-anointed several times in succession. Then he wandered off, licked the tiled hearth and gave another performance there. After that, he self-anointed after lick-ing the leg of a chair, the handle of a coal shovel and a number of other items of household furniture, wandering from one room to another.

It was a busy time, especially for Jane, moving floodlights and manoeuvring the camera, with Mr and Mrs Castleman always patiently attentive but never getting in the way, as can so easily happen when one is filming other people's pets and they are over-anxious to help. During the first session of self-anointing that afternoon, when Rufus used the carpet, he alternated biting the material of the carpet with licking it, quite marring the look of it. Neither our host nor hostess showed any sign of annoyance. Finally Rufus fetched up at the side of one of Mr Castleman's leather sandals. He started again to lick and self-anoint.

'I wonder' said Mr Castleman, 'whether human sweat is the attraction, for example, on the handle of the shovel and so on.'

There was a pause while we all pondered this.

'We can test it,' said Mr Castleman. 'I received a new pair of sandals by post this morning and I haven't undone the parcel.'

He fetched the parcel, unwrapped the sandals and, taking care not to touch the leather, placed one of them on the ground. Rufus wandered over to it and started to lick its polished surface, near the toecap. This time, also, the hedgehog alternated licking with biting and in a few minutes had scraped the polish off the sandal, scarring the surface down to the bare leather for the length of an inch by nearly half-an-inch. To his credit Mr Castleman made no attempt to stop the experiment.

The afternoon ended with Rufus licking and biting at the base of a bookcase, again leaving an ugly scar on the otherwise unblemished surface, after a continuous session of self-anointing lasting a full twenty minutes. Mrs Castleman made no protest either. By this time there were four weary humans and one active hedgehog prepared to continue. Rufus wandered to another part of the house. Mrs Castleman invited us to take tea.

During this same period, many people wrote to me or told me about hedgehogs they had seen self-anointing and of the objects or substances licked. The list of these is almost endless. One of these accounts stands out in my memory. The lady who told it said how she had a pet hedgehog and was anxious to find it a mate. One day she met a gipsy carrying a hedgehog.

'Where are you going with that hedgehog?' she asked.

'I am going to the fair, and I hope to sell it,' replied the gipsy.

'How much do you want for it?'

'Half-a-crown.'

She bought the hedgehog and told the gipsy why she wanted it.

'You put the two together, when you get home,' said the gipsy. 'If this one licks your hedgehog's spines and then puts its spittle on its own spines, this means they are a pair and will mate.'

'How do you know this?' she asked.

'Gipsies have always known it,' he replied.

So it seems that gipsies knew about self-anointing long before scientists did.

Since I first drew attention to it in print many other people have witnessed self-anointing and have photographed it, so that today the self-anointing of hedgehogs can be accepted as a commonplace phenomenon. Several explanations have been put forward to account for self-anointing, although none is fully satisfactory. It has been suggested that it may be part of love-making, as the gipsy suggested, a defence against enemies by masking the animals' odour, a displacement activity, an addiction. More of this later.

Soon my attention was focused on another hedgehog mystery. From at least as far back as classical times there have been stories of hedgehogs rolling on apples and carrying them away on their spines. They have been said to do the same with pears, grapes and strawberries. There are people who claim to have witnessed them in the act and many more who tell stories of circumstantial evidence pointing to the truth of it. But nobody believes them.

I have interviewed some of the people who made these claims, or have exchanged correspondence with them, and I found it hard to set the idea aside.

Two arguments are commonly used by zoologists to refute the idea. The first is that hedgehogs eat slugs, worms, insects and the like, never fruit. I remember, at the time I was first keeping hedgehogs in captivity, hearing Frances Pitt, a leading naturalist in her time, broadcasting on the radio. I believe she was answering a question sent in by a listener. To the best of my recollection she said: 'Hedgehogs never eat vegetable matter of any kind.' (She had, of course, overlooked for the moment that any hedgehog will greedily devour *bread* and milk.)

She then continued: 'As for their eating fruit, this is sheer nonsense.' And, metaphorically speaking, she fairly split her sides laughing at the idea.

It so happened that we had been trying our various hedgehogs with fruit. One ate banana but would touch no other fruit. One only ate apple, another ate pear, and so on. But all ate fruit of one kind or another.

The other argument against the idea of hedgehogs impaling apples is that their spines would not pierce apple skin. People had tried dropping an apple from a height of two feet onto a hedgehog's back and found that the apple simply rolled off. This does not surprise me. There is a world of difference between a natural situation and one that is contrived. If, for example, somebody tried to press a drawing pin into the sole of my foot I would make it difficult for the operation to succeed; but if I tread on a drawing pin on the floor, while wearing slippers, the point is driven home.

It occurred to me, soon after we had filmed the baby hedgehog, that if a hedgehog happened to self-anoint where windfall apples lay on the ground it might conceivably spear one or two during its contortions. When self-anointing, the hedgehog often raises one side of the body and presses the other towards the ground. Should it do this where there were several apples lying, I argued, it might well press its spines into one or more of them.

To test this we took a hedgehog into the woods when crab apples had begun to fall – I wanted to make the test under completely natural conditions – and Jane was with me to photograph it. We put the hedgehog on the ground. It started to sniff around, as hedgehogs will. Then it found a dead leaf soiled with animal droppings. It started to lick this. Then it self-anointed, and in no time, as it contorted itself, it speared an apple and walked away with it on its spines.

Every naturalist to whom I have shown the photograph has either said frankly he did not believe it or we have seen by his expression that he had his doubts. My good friend, the late Maxwell Knight, said: 'Maurice, I will believe it when you show me a film of it.' But it is just as easy to fake a film as a still photograph. Perhaps we ought to have staged the same thing in a studio, under the floodlights. That, for me, would have been faking it, and I argued that surely if I gave my word that this thing really happened people would accept it. Alas, the answer is 'No'. So I gave it up and turned to the many more interesting things I had on hand. I have always regretted that I did not persevere with this to obtain a film record – and, secretly, I still believe people who claim to have seen hedgehogs carrying away fruit in this manner.

Another baffling piece of hedgehog behaviour came to my notice one summer's evening, when we were living in the house at Horsley. Just as it was getting dark, I heard an agonized cry coming from somewhere across the road. It was repeated every few seconds. Clearly, some animal was in distress, but what it could be I had no idea. Opposite our house a track led into the woods and, to the right of this, began a line of houses each with a long garden that ran back to end in the woods. The gardens were all plentifully planted with shrubs and well-grown trees. So, as I stood looking across the road, listening to the agonized cries, I was looking into a mass of vegetation in which each house was half-submerged. And the light was fading rapidly.

I walked across the road and along the track into the woods, straining my ears to pinpoint the source of the cries. Clearly they were coming from one of the gardens, so I turned right. This took me into boggy ground covered with scrub and thorny bushes, and so to the boundary of the first in the row of back gardens. The repeated cries told me I was heading in the right direction, but as I had not had the wisdom to take a torch with me I was less informed than I might have been about the hazards that separated me from the animal crying out in pain.

The first hazard was a ditch, fortunately dry at that time of the year, at the foot of a tall, think unkempt hedge. I fell into it. Having picked myself up I found that passage through the hedge was made less than easy by strands of rusty barbed wire lying about among the stems making up the hedge. And I had sprained my ankle! Nevertheless, come what may, I was determined to reach the source of the cry, if only to solve a mystery and add to my knowledge. So through the hedge I went, largely by dint of brute force and disregarding scratches, and across the first garden, with the cries sounding louder at each step.

The second hedge was slightly less formidable and soon I was in the second garden. The cries were now near at hand, harsh, heart-rending. In the gloom, and partly by touch, I identified an enclosure bounded by six-foot wire-netting with a hen-house at one end. In the less than half-light I could make out a hedgehog, caught by a hindleg in a steel gin-trap. As I watched, I saw the hedgehog tug to free itself of the remorseless steel jaws, which were firmly closed on its leg. While it tugged it was silent. Then it would abandon the tugging and raise its head, opening its mouth and giving vent to this unearthly screaming. It could hardly have been a cry of pain, because the worst pain must have been while the animal was tugging to free itself. Here, then, was a first-class mystery, one I have often pondered since that nightmarish evening.

I could do nothing but stand and stare into the hen-run for, as I reasoned then, I could only have reached the hedgehog by tearing down the wire-netting with my bare hands. This would have been a gross infringement of private property, even if I had been able to wrench the netting from its posts. There was nothing for it but to return the way I had come, go to the house at the other end of the garden and tell the owner of the house what was happening and hope he would allow me to do something about it.

I retraced my steps and at last reached the gate leading to the front door of the house, hobbling painfully on a throbbing ankle. My sympathies were by then even more with the trapped hedgehog! The only lights in the house were at an upstairs window. The family was going to bed. They could not have failed to hear the screams but had chosen to ignore them. The trap had probably been set deliberately to catch the hedgehog coming to steal eggs. So clearly it would have been a hard, if not impossible, task to persuade them to effect a rescue, and they would not have thanked me for my trouble. With my conscience heavy, and still nursing my injured ankle, and with the cries of the trapped animal ringing in my ears, coward that I was, I went to bed and pulled the blankets over my head, pondering mystery number one.

A hedgehog is a singularly silent animal, vocally, yet by virtue of its spines it is remarkably noisy. You hear what sounds like a panther crashing through the bushes only to see a diminutive hedgehog emerge at the end of it. Many a time people have thought there was a burglar prowling around the house at night and have called the police. The intruder has proved to be a hedgehog rubbing its spines on a concrete path, or one that has got into the house rubbing its spines on the skirting. Usually, so far as its vocal cords are concerned, it is silent, giving out at most low snufflings, snores or twitters.

Are its cries, when trapped, cries for help? Hedgehogs are solitary animals, hardly given, one would have thought, to calling to their fellows for assistance, nor could these render help even supposing they responded. Is the screaming a mechanism implanted in each individual and designed to alert their fellows to potential danger? Is it a cry of defiance at some enemy unseen or not fully comprehended? The only other time I have heard this scream was in broad daylight, when a hedgehog suddenly found itself face to face with Jason, the mastiff-like boxer-cross we used to have. David hedgehog screamed at Goliath dog, as if in defiance, instead of rolling into a ball of prickles, as it is supposed to do at the first sign of danger. Was the screaming of the trapped hedgehog no more than a scream of defiance at an unseen menace?

The screaming might also be interpreted as an infantile alarm call intended to alert its parent to its distress, as a mortally wounded soldier on the battlefield or a woman in labour have been known to cry 'Mother'. This theory is discounted by watching, as we often did, the two baby hedgehogs Mrs Fulbrook had brought. They used only a low, bird-like whistle when they had wandered away from their mother and got lost.

It has often been said that a common trick used by country boys in times past was to rub a stick across the hamstring of a hedgehog to make it scream. The hedgehog I saw in the gin-trap had the teeth of the trap across its hamstring! I have dismissed the scream as a cry of pain mainly because, as I have said, the scream came when the pain must have been less intense. But why should the animal scream when pressure is applied to its hamstring?

With these musings I fell asleep, my head still under the blankets. And I have never caught up with this unsolved problem.

Hedgehogs have the reputation of being egg-thieves. This is why, in pheasant preserves, they were shot or clubbed without mercy and their grisly carcasses hung, with those of rats, stoats, weasels and hawks, on the gamekeepers' gibbets that used to be so common a feature of the English countryside, and elsewhere. And here is another mystery. Many people have, like myself, put eggs down near a hedgehog to test whether it will eat them. Almost invariably the animal has ignored them. We tested our hedgehogs and of the dozens we have had none has ever eaten an egg made available to it. Yet careful study by zoologists during the past twenty years has shown that some hedgehogs will habitually take the eggs of ground-nesting birds. They are also known to take the eggs of domestic poultry, which was doubtless why the gin-trap was set in the hen-run I had been at pains to locate. The truth is, probably, that some learn to eat eggs but that most hedgehogs have not the faintest idea of breaking the shell to get at the contents.

A man once wrote to tell me how he had caught a hedgehog in

the act of consuming an egg in his hen-house. What was more remarkable was that he found that the egg had been pressed into the compacted hay of the nesting-box, and there it stood on end as if in an egg-cup; and the hedgehog was eating it. Being a humane person, he was unwilling to kill the rogue. He merely removed the hedgehog and set it free at a distance from the hen-run. The next evening he again found a hedgehog in the hen-run, presumably the same one for it also was using an 'egg-cup' in the hay. This time he put a dab of red paint on the animal's spines, put it in his car, drove a quarter of a mile and released it. The next evening the animal, complete with red paint, was back in the hen-house using another egg-cup. It was again put in the car, driven away and liberated half a mile from the hen-house. The following evening it was back; this time it was taken a mile away. It did not return. Aside from the egg-stealing, this was an interesting experiment on navigation.

Generally inoffensive, an endearing pet, a friend of the farmer because it feeds largely on small agricultural vermin, the hedgehog has another fault, that of milk-stealing. Its passion for bread-and-milk needs no emphasis. Nevertheless the belief among cowmen, over centuries, that hedgehogs will sometimes take milk from cows in pasture has been pooh-poohed by scientists. When, for example, a cow taken to stable for milking has been found to be dry, generations of cowmen have remarked: 'A hedgehog has been at her.' This has, however, been treated as a legend and has been met with scepticism among zoologists. As is sometimes the case, the scientist has performed lengthy circumlocutions to explain away what he considers to be a legend, instead of trying to test it. He has suggested that possibly a hedgehog has been rooting under the cow's udder for insects, that have crept there for the warmth, as the cow rested on the ground. Then, if the cow happens to have a leaky teat the hedgehog has lapped up the milk and the rest has been imagined. The stock argument has been that the hedgehog's mouth is too small to take a cow's teat; and in any event, what cow is going to let a hedgehog take its teat in its mouth without protest! And would not the hedgehog bite the teat? So the seemingly convincing arguments have run, countering the assertions of farmers and cowmen.

I had spoken to people from time to time who claim to have seen a hedgehog at work stealing milk and I felt convinced, from the way they spoke, that there must be some truth in the old

belief.* One farmer claimed that he had come across a hedgehog
one morning near one of his cows. He happened to have his gun
with him, so he shot the hedgehog, ripped it open and found its
stomach full of milk. None of his cows had a leaky teat. Assuming
the farmer told the truth, and I had no reason to suppose otherwise, what greater proof is needed? In any event, a leaky teat does
not spill that amount of milk, except over a long period.

As a result of speaking to these people I did what any orthodox
researcher does when faced with a similar problem. I formulated a
theory and then set about testing it. Hedgehogs have been seen by
none other than the authority on the species, Konrad Herter, to
stand erect on their hindlegs. If a hedgehog could do this would it
not comfortably reach the teats of even a standing cow? Moreover,
if the old argument, first put forward in print by Barrett-Hamilton
in 1910, that the animal's mouth ws too small to take a cow's teat,
could be shown to be fallacious, that would be the first hurdle
surmounted.

As I have said, Jane took a film sequence of Rufus biting at the
wooden base of a bookcase. It was possible to measure these
tooth-marks and to show that Rufus had a gape of not less than
1¼ in., measured at the front of the mouth. And I reckoned that
if a full-grown hedgehog stood erect on its hindfeet the tip of its
nose would be not less than a foot from the ground. So I took a
foot-rule and went forth into the fields.

In the usual course of events, anyone seen going from one cow
to another in a field, holding a piece of wood under their udders
(even if it is a foot-rule) is likely to attract attention. So I chose
my moment carefully to carry out these measurements; but I
became so engrossed that I did not notice the approach of a cowman.

'What do you think you are doing?'

The loud voice, interrupting my quiet studies, startled me. I
swung round to face a youngish man looking at me quizzically,
more in sorrow than in anger, I thought. At least he looked puzzled.

'Good morning,' I said.

Courtesy costs nothing and it gives one time to think.

'I'm doing some homework,' I continued.

---

* I have already dealt with this subject in my book, *The Hedgehog*, published in
1969, but it does no harm to summarize here some of the points made in that
book and to amplify a few of the details.

The cowman still looked puzzled so I thought it better to explain.

All the milking cows I had measured had teats that reached well within a foot of the ground. The diameter of the teats averaged less than 1¼ in., so it was physically possible for a hedgehog to take the teat of a standing cow into its maw. Domestic cows are patient animals and seem to be resigned to having their teats touched by all and sundry: I recall a photograph in one of my files of a cow suckling several piglets.

The cowman seemed reasonably assured that I was not wholly insane. Indeed, he became quite friendly and told me about the cows he had had that had gone dry.

With one consideration after another, I built up my hypothetical model supported by numerous hypothetical arguments, but I did not have the wit to do what my good friend Mrs Stéphanie Ryder did. After we had been discussing this matter she obtained a calf's sucking teat, the modern counterpart of the feeding bottle, filled it with milk and collected a hedgehog. The hedgehog co-operated fully. It sucked the contents of the artificial udder to the last drop, and needed little prompting to do so. Providentially Mrs Ryder received a copy of the current *Veterinary Journal* very soon afterwards, which contained photographs of cows' teats with lacerations consistent with injury from the teeth of hedgehogs.

Normally a cow gives more milk in the morning than in the evening. Where the yield is reversed something must be amiss, and when a cow is dry in the morning and gives milk in the evening suspicion turns to our spiny friend. And if a hedgehog can milk a cow dry it has proportionally something of the liquid intake of a camel.

When I first began to study hedgehogs I conceived the light-hearted theory that these animals are basically desert animals that came from the tropics and spread into the temperate regions. Facetiously, I went on to suggest that, blessed with spines, they were like the desert plants, including cacti. Their drinking habits also seem to support this. A camel has been known to drink thirty gallons in ten minutes. A donkey, another desert animal, can rival this performance. I have a record of a hedgehog drinking a pint of water straight off, and for its size this must equal a camel's performance. Moreover, a hedgehog hibernates; and a few years ago one scientist suggested that hibernating animals are descended from tropical animals that have spread into the temperate regions.

Another light-hearted theory I would now put forward is that a hedgehog is prone to addictions. Some become addicted to eating eggs, some to stealing milk, others to self-anointing. So let us return to the possible explanation for this last intriguing behaviour. It may be part of love-making, as some claim. It may be to disguise the natural body odour, as a defence against predators. If so, what are we to say of the hedgehog that found a cooked beetroot on a garbage heap, licked it and self-anointed vigorously and finally went away coloured a rich purple instead of the natural drab colouring. Are we to suggest it was changing its colour as a protection?

When I was discussing this subject with a naturalist friend she told me how, when she was a child, she was forever copying the noises and movements of the animals she saw. Among others, she imitated a cat washing itself. She soon found that if she licked her hands, arms and legs, copying the cat, the smell of her limbs was strong and, to her, pleasant once the saliva had dried. On the basis of this, she was of the opinion that a hedgehog putting spittle on its spines would greatly strengthen its natural odour rather than mask it – which, to me, sounds eminently logical.

It may well be that each of these theories has some truth in it. Even so, it still does not rule out the possibility that each of the processes suggested can be taken to extreme excess, and habitually so, which is the essence of an addiction. It is even possible that there may be another feasible explanation, a parallel with the kangaroo's 'water bag'. A kangaroo is said to lick its wrists in the heat of the day, the evaporation of the saliva serving to lower its body temperature. On the same principle, some people drink a cup of hot tea, which promotes sweating, and the evaporation of the sweat lowers the body temperature; tea can be cooling on a hot day. The evaporation of the saliva from self-anointing may conceivably be a cooling agent, of use in a desert animal when it happens to be out at midday, and perverted in its descendants that have come to live permanently in cooler latitudes.

Whatever the explanation, the sight of a hedgehog self-anointing is, to use a modern ill-used idiom 'quite fantastic'. Thus, in 1977, a lady asked our help. She had found a family of motherless baby hedgehogs. She had hand-reared them and now, as adults, they were her pets. She had to go to look after her mother who was sick; but she could not take her pets with her, nor could she find a safe place in the built-up area in which to release them. We agreed to take care of them, happy in the thought that in a month's time they would hibernate and would then be no trouble to look after.

It was easy enough to adapt an empty aviary for their reception; there was sufficient floor space for them to wander. All that was needed was a heap of hay at one end in which they could make their sleeping nests. In any event the agreement was that we would house them for a few weeks before they went to sleep for the winter. Then, when they woke again in spring, we would liberate them in some place where they were unlikely to be at risk from natural predators or heartless humans.

They were duly brought to us and installed in their new quarters. Within minutes of their having been placed in the aviary, and without anyone noticing how it all began, we found all five grouped in the centre of the aviary floor, with only a little space separating one from another, and each was self-anointing. It is a bizarre sight to see one hedgehog doing this, but five performing simultaneously beggars description. Imagine five balls of prickles, within an area of about a yard square, licking and foaming at the mouth and alternately throwing themselves into contortions – a circus act in miniature.

We paid for this spectacle. The hedgehogs should have gone into winter-sleep in mid October. If you keep these animals in captivity there usually comes a time when the food you put in for them one evening is found untouched the following morning. This did not happen with our five. Day after day the food bowl was emptied and had to be replenished. In vain we looked for the tell-tale signs of hibernation. It was costly and time-consuming feeding them throughout the winter. The only consolation was that we obtained some evidence that those who assert that hibernation is a habit which protects animals against a potential shortage of food are probably near the truth.

9 *Jasper, the jay*

It was spring, in 1955. For the past few days I had been trying to hand-rear a baby bird but, despite my best efforts, it had died. One of the drawbacks to hospitalizing animals is that when you fail remorse hangs like a pall over your head for the next twenty-four hours. I always remember being surprised when a zoologist colleague, not noted for his soft feelings about animals, confessed to me that if ever a baby bird that he was trying to succour died on his hands he felt depressed for the next day or two. The best antidote to such depression, as I have found, is to be presented with another patient to tend.

On this beautiful spring morning, feeling dejected by my recent failure, I had walked out of the house and was standing on the doorstep, hands in pockets, gloomily turning over in my mind what else could have been done to keep the spark of life going in the pathetic corpse I had left behind indoors. Then the front gate swung open and two rough-looking characters came through the gateway. One was carrying a large bird's nest in his cupped hands.

'We were cutting down a tree,' he announced, 'and when we came to strip the branches from it we found this' – indicating the nest with two fledgling jays in it. 'We supposed the parents would have nothing to do with them, even if we put the nest into another tree, so we thought we'd bring it to you.'

I had not met either of the two men before, but presumably our reputation for accepting abandoned birds had reached them.

Gone was the gloom. All my attention was now on the nest and its occupants. The two nestlings were almost completely bare, their bodies grotesquely egg-shaped with absurdly large legs and long skinny necks that stretched up to support insistently gaping mouths. The one certain thing was that the young jays were hungry. It was easy enough to feed them. One had only to push food down their throats as they thrust their wide-open beaks upwards for more. Unfortunately, one of the nestlings died a few days later. But the second, after an initial period of doubt, gained strength and prospered.

Corbie, the rook, when presented with a match always held it under his foot and pecked the match-head vigorously. When the match was a self-striking one it burst into flames

Ermintrude, the tame stoat, delighted in digging out the earth containing favourite pot plants

Stoats are adept at dribbling an egg along the ground. In the wild they have often been observed stealing eggs in this fashion, or holding an egg under the chin by the forepaws

Niger, the rook, behaving like a Phoenix as the flames from a handful of burning straw rise in front of him. In spite of his obsession with fire he never seriously hurt himself

Niger, the rook, having struck the head of a self-striking match with his beak is about to pick up the lighted match to hold under his wing

A study of self-anointing, the puzzling trick used by the European hedgehog in which the animal coats itself with its own saliva

A hedgehog self-anointing, apparently throwing itself into convulsions and foaming at the mouth, a commonplace piece of behaviour which eluded observation by zoologists until fairly recent years

Jasper, the jay, in the act of anting, a puzzling trick seen in many songbirds

The common jay, when anting, brings its wings forward and arches them as if forming a canopy

The common rat has less of a reputation as a climber than the ship or black rat but made no bones about climbing this metal rod

A pair of mallard resting, the cryptic plumage of the hen or duck contrasting with the bright colours of the drake

A corner of the garden at Weston House showing a cage in which a ferret was living and ducks wandering at will on the lawns

At this stage of its career food and sleep were its only require-
ments. As it grew in size and its feathers began to appear, it began
to develop a personality – it was no longer just 'a jay' or even 'the
jay' – and this was symbolized in the development of its plumage.
At first both personality and plumage were non-existent, then
both passed to the stage of being nondescript, and finally there
was a recognizable orderliness. So far as the plumage was concerned, the orderliness could be epitomized in the way the patch
on the side of each wing, of blue feathers with their black stripes,
changed from disarray to a neat design, setting the seal on this
colourful bird with its pink-brown body, black tail and white
rump.

A personality must be given a name and the jay now became
Jasper, at Jane's suggestion, after Count Jasper, the traditional
villain of Victorian melodrama, who came on stage stroking his
long black moustache and muttering 'Aha!' (The common European jay has black moustachial stripes running back from the
base of its beak.)

In the early stages of a young animal's life, physical and
psychological changes clearly depend on a changing internal
biochemistry and owe little to the influence of the outside world.
In due course, however, the environment begins to have its
impact: it is almost impossible to describe in words, but we can
exemplify the changes in a jay's personality by a single factor –
the use of the voice.

At the very beginning, as a nestling, the pangs of hunger produced in our jay nothing more than a readiness to push its gaping
mouth on its scraggy neck towards the heavens the moment the
nest was touched. It was the usual response of a young bird to a
parent returning to and landing on the nest with food. The time
came when the gape was accompanied by an insistent and high-
pitched squawking, and this continued, long after the fledgling
stage, at the mere sight of food or even at the sight of someone
carrying the food bowl. The squawking was obviously inherited. It
had a physiological basis and a functional value.

After Jasper had started to feed himself he grew relatively silent.
Almost the only sound we heard from him was the usual raucous
and deeper squawk typical of the European jay, a harsh scolding
cry, emitted in moments of panic, discomfort and distress. The
sight of a cat too near the aviary, for example, would call forth this
typical jay squawk. There were also times when it appeared to be
prompted by sheer *joie de vivre*. The sounds that interested me most,

however, were those he made which owed little to inheritance and which, it can be argued, may justifiably be given the title of an art.

Thus, if we try to follow the sequence of events resulting in the high-pitched squawking of the nestling or the deeper squawk of the adult, we have something of the following. In the squawk generated by hunger the sight of food operating through the eye, or a stomach reflex operating from within, sets in motion a train of nervous and muscular reactions. These control the outlet of air across the vocal cords and trim these cords appropriately to produce sounds common to all young jays. In the harsher call of the adult a stimulus received through the eye sets up a state of physiological disturbance the visible symptoms of which we describe as fear, alarm, or what you will. The physiological effects set in train the nervous and muscular reactions controlling the expulsion of air and the movement of the vocal cords appropriately produces a sound common to all subadult or adult jays.

These two processes have the following features in common. The sounds are made in the same ways by all jays of that species, whether the birds grow up among their own kind or, like Jasper, are reared in isolation. They are therefore not learned but innate or inherited. The pattern of nervous and muscular response is there before hatching. Both processes are basically physiological and they indicate hunger, fear, or perhaps well-being. Both are functional except perhaps when the raucous call is indicative of well-being. They are their native calls.

But there came a time when we heard calls issuing from Jasper's aviary, sufficiently strange to lure us out of the house to see what unusual visitor was there. Instead we found that the jay was mimicking the call of some other bird. Soon he was mimicking perfectly a number of the wild birds visiting the garden. On one occasion he was making a sound we could not identify until we suddenly realized it was the laugh peculiar to someone then visiting us regularly! He also mimicked the guinea-pigs in their nearby pen and he mimicked Robert imitating the call of the guinea-pigs — and the call of the guinea-pigs and Robert's imitation were quite distinguishable when Jasper copied them.

There is nothing new in all this. Many birds have the ability to mimic either mechanical sounds, natural sounds or human speech. Parrots are the best known but the practice is more widespread among birds than we normally suppose. The only reason why Jasper's efforts are given prominence here is because I was able to observe him closely; and it is from these observations that I feel justified in arguing the possession by birds of an artistic streak, however simple or rudimentary that may be.

It is never easy to say precisely what we mean by an art. One definition which seems to have much to recommend it is that a craft is functional and an art is non-functional. A potter may throw a jar the purpose of which is wholly utilitarian. The shape may be pleasing to the eye, but that is secondary. The original purpose is utilitarian. If the potter decorates the jar in any way, however, the decoration adds nothing to the usefulness of the jar. It is non-functional and in the decoration we see the beginning of an art, something pleasing to the senses of potter and purchaser alike.

I have emphasized the process by which innate calls are produced because I want to compare it with the process of mimicking. If you stood in front of Jasper's aviary and whistled in a particular way he would give you his attention and no more, watching you intently, perhaps with his head on one side. When you departed he would hop about the aviary as usual and then, some time later, you would hear him copying your whistle perfectly. One has the impression that the original sound of the whistling has been received through the jay's ear, been registered as a hearing memory in the brain and is later, through neuro-muscular control, reproduced perfectly by the vocal cords. The process is very similar to that seen in a tape-recorder but with a significant difference.

A sound imprinted on a recording tape will be reproduced when the appropriate mechanism is set in motion, whatever the circumstances. It is a purely mechanical reproduction. A bird will utter mimicking sounds only when it is in good health, when there are no disturbing influences around: in short, when it is feeling on top of the world. In other words it will, usually, use these calls which it has learned only when it is in a relaxed mood and enjoying life. To that extent it may be said to be using the calls for the sheer pleasure of doing so. This, it seems to me, is art in its simplest form.

The crow family, of which jays are members, is generally accepted to include the most intelligent of birds. Intelligence is a most difficult quality to define or assess but, this qualification apart, we all have a fair idea what is meant by the word although there may be differences of opinion as to which birds have the strongest claim to possessing it. At all events, with Jasper we had the opportunity of studying it in at least one individual.

At the fledgling stage Jasper was put in the same aviary as our young rook, Niger. By that time we had a dozen different species of birds in the aviaries, comprising twenty-five individuals, most of them hand-reared and all under constant observation. Of these, the most versatile in behaviour and the most rewarding to watch were the rook, crow and jay; and the one giving the appearance of greatest intelligence was the jay. Although Niger was at an age when he should have been capable of picking up food for himself he still insisted on being hand-fed. To save effort on our part we fed him, at the same time as Jasper, through the wire-netting of the aviary, holding the food through the netting with forceps. It was noticeable that Jasper, of the two, learned more quickly to respond to the position of the forceps. At first, for example, when one of us appeared with a bowl of food, both the rook and the jay would start to call and to gape without making an effort to come to the netting. It was the jay that learned to do so first.

A number of branches were fixed around the inside of the aviary, with their twigs towards the netting. I formed the habit of tapping the wire with the forceps at the point where I wanted the bird to alight in order to feed it. The first time, the jay responded at the third tap. Thereafter he would come at the second or the first tap, or even if I rested the tips of the forceps on the wire. Niger took much longer to learn the significance of the tapping, so long, in fact, that he appeared to do so in the end by imitating the jay. Despite this, both often perched too far in from the wires so that the food held in the forceps hardly reached their beaks. It was noticeable, however, that the jay did this less often than the rook and was the more ready to adjust his position for the forceps to reach him.

During the next few weeks, before Niger was transferred to his permanent aviary, we gained the impression that Jasper had the more impressive personality. This may have been more apparent than real, in part perhaps because he was quicker in his movements; he also had the more expressive face due to the movements in his crest and eyes. The crest was erected the moment we spoke to him and this had the same effect on an observer as seeing a person raise his eyebrows, as if trying to understand the meaning of your words. The alternate raising and lowering of the crest imparted a mobility, almost a vivacity, to the jay's face.

It was while watching his face closely that I became aware of a similar mobility in his eyes. The feathers surrounding them were frequently being moved, the pupils contracted and expanded and, above all, we found the eyes were capable of being moved independently of each other, like the eyes of a chameleon, but more rapidly. Indeed the movement was so rapid that we could not have been sure of it had Jasper not had the incurable habit of staring one straight in the face at close quarters. In addition, he allowed me to hold his beak to keep his head steady while I watched his eyes. It must be confessed that I had never before had reason to suppose that any bird possessed this chameleon-like trait. Supposing that this represented a gap in my education I made tentative inquiries of ornithologist friends, but none of them seemed to have met it or heard of it.

I started to look around to see what our other eleven species of birds could do. The tawny owls could be eliminated at once. Pigeons I found to have eyes that move in concert. Corbie and Niger, the two rooks, both gave me no opportunity: try as I might, they would insist on turning their heads to look at me with one eye. The rest gave no help either. I examined a number of birds in the London Zoo and tried to watch others in the wild but it seems the jay's ability to move the eyes in this way must be rare.

The value of this can be exemplified by the occasion when, while Jane was talking to Jasper, our dog came up to her. The jay cocked his head slightly and with one eye looked obliquely downwards at the dog and with the other eye looked horizontally forwards at Jane. Presumably he was able to focus both eyes simultaneously. This leaves one wondering about the mechanism involved, and whether a jay can use its eyes in somewhat the same manner as we can use our sense of hearing. I can listen to a symphony concert on the radio, while writing this, for example, enjoying the music but partially and not giving my whole attention to the writing. At any moment I can put down my pen and enjoy the music to the full or I can shut out the music and pay more attention to what I am writing.

It would also seem that a jay has something approaching binocular vision for Jasper would look one straight in the face, his eyes situated neither to the side nor at the front of the face but both directed at the same angle to one's face.

In making these studies of Jasper's use of his eyes I had my face close to his and he was quite unafraid of this, so I experimented with putting my hand near him at the same time. He was slightly shyer of this and sometimes moved away. On one occasion he watched intently as I moved my hand slowly towards him. When it was almost level with his perch he quickly hopped out of reach. It was not my hand that had caused his retreat. It was a tiny insect six feet away along the perch, so small that I could not see it until he returned holding it in the tip of his beak. Moving his eyes independently he had watched my hand with his right eye, at the same time marking down the insect with his left. The combination of acute vision and the ability to give close attention to two different events simultaneously, even to pin-pointing an insect smaller than a pin's head six feet away, must confer great advantages.

Granted that the use he made of his eyes and the movements in the feathers on his head combined to enhance his personality, it was nevertheless his skill at vocal mimicry which made the greater contribution. During his first year Jasper learned to imitate the calls of the mallard, owls, cock pheasant and pigeons we kept. He also copied the sounds of our foxes at play. He also reproduced perfectly the songs of a number of wild birds frequenting the garden. It may have been the milkman who encouraged Jasper in his vocal mimicry. Early in the morning he came up the drive whistling, always the same tune. When he reached Jasper's aviary he would stop and talk to him. Soon Jasper was whistling the milkman's tune, talking in his voice and making the sound of milk bottles clinking together.

Once started on the road as a mimic he was quick to add to his repertoire. Jane used to go round each morning with water for the aviaries, carrying a watering can in one hand and a pail with a squeaky handle in the other. Jasper copied the sound of the squeaky handle. Some days later Jane used a pail that did not squeak. Jasper supplied the squeak. After that any galvanized pail set him squeaking.

At that time we kept some of the animals' food in two small galvanized iron bins, cylindrical and two feet high. These were portable and anyone carrying one within sight of Jasper was greeted with the sound of the squeaky handle, a clear association of similar objects with the one sound.

Jasper soon learned to call our dog Jason by name. Jane used to take the dog out for a walk every morning. On their return Jason, reluctant to come home again, would dawdle two or three yards behind her. She would turn and say sharply 'Come on, Jason' or 'Come here, Jason'. Before long, whenever Jason went near his aviary, Jasper would look down at the dog, head on one side, and scold 'Come here, Jason' in Jane's voice.

All this is commonplace, yet remarkable when one reflects on it. One would expect the bird to associate the sound of a squeaky handle with a pail, or anything resembling it, because the sound comes from the object. In the Jane-Jason context the dog is perceptibly removed from the sound. It seems reasonable to say that the jay appreciated that Jane was calling the dog by name, for although the sound of the dog's name was coming from a source several feet from the object, the bird related the two.

Another point of interest arose when a very noisy motor-cycle went past the house. We had an ornamental brass knocker on the front door shaped like a piece of tree trunk with a woodpecker seated on it. By turning a knob near the base the bird was made to peck at the trunk, giving out a loud metallic rattle. The knocker was seldom used but that did not prevent Jasper from learning to imitate it. As the motor-cycle was going by it suddenly sounded as if its motor had gone wrong. It was Jasper making the door-knocker imitation, using the nearest sound in his repertoire to mimic the noise of the motor-cycle. For Jasper's mimicking to sound like a motor-cycle illustrates how loud was the sound made by the jay in imitating the door-knocker.

There was another remarkable sequence. Our neighbour had a small daughter, Sally, who, when her father left for work in the morning, followed him down the drive calling 'Bye-bye'. One could see only bits of Sally through gaps in the dense vegetation. We can be sure that Jasper never saw the whole of Sally together in one piece until one day her mother brought her to see us. As soon as she appeared at the gate Jasper greeted her with a rollicking string of 'Bye-bye'. There seems little doubt he had recognized Sally from the bits of her he had seen through the hedge, fitting the bits together in his brain like a jig-saw.

Jasper's 'talking' seemed to expand his personality and this and his extreme tameness made him a great favourite of us all. We could almost say we were close friends. This is not too extravagant a phrase, for whenever I went into his aviary, as was necessary to supply his food and to renew his bath and drinking water, he would fly immediately onto my shoulder or my head. There he would perch as long as I could spare time to allow it, with no suggestion of fear or aggression on his part.

This relationship was maintained until the second spring after Jasper had first joined us. Then the territorial instinct showed itself.

In that spring and in every succeeding spring, at the beginning of the breeding season, Jasper treated anyone entering his aviary as an intruder. I noticed this for the first time when, as soon as I entered, he whistled in the sweetest possible tones. At the same time he crouched on his perch ominously. Then he flew over my head, striking downward with his feet at my scalp, and landed on another perch farther away. My head was painful for a while, sufficient to put me on my guard the next time I went in. The whistling was subdued, a subsong, and it was recognizable as the milkman's whistle, being used as a threat.

At the same time I noticed that if I did no more than walk around the aviary, on the outside, he would follow me, on the inside of the netting, showing clear signs of aggressive intent. I was, in fact, walking round the boundary of his territory and his purpose was to see that I did not cross it. In an effort to allay his fears, to soften his aggressiveness, I put my face near the wire-netting to say to the bird, who was the other side of it, 'Hello, Jasper', using my gentlest and most placatory tones. Very soon I found that, when I walked around the aviary, the jay was keeping abreast of me, saying all the while, in perfect imitation of my own voice: 'Hello, Jasper. Hello, Jasper.' After that, whenever I went into the aviary and heard what sounded like myself using these words, I knew the jay was about to attack.

There is nothing new in this. It is well known that a wild jay harassing an owl will hoot at it. It will caw like a crow that is flying overhead. Jasper would miaow aggressively at our cat if it went near the aviary. There seems to be some comparison here with the way human beings, children especially, when aggressive will copy the voice of the person they are attacking or threatening, even using that person's words, in a derisory way.

In the third year we were asked to take into care a tame hen jay, promptly named Mrs Jasper by Jane. It seemed an ideal opportunity to give Jasper a mate, but although he accepted her as an aviary-companion it was with no more than a thinly-veiled tolerance. They never mated, although Jasper made the beginnings of a nest, and the only time they showed any unity of purpose was when both called in the harsh scolding notes natural to jays when a stranger came through the gate or a strange cat walked towards the aviary, or even when a large bird such as a heron or a crow flew by overhead.

We logged some sixty separate mimickings by Jasper, of mechanical sounds, speech and the songs of other birds. Mrs Jasper also proved to be an accomplished mimic but she used her repertoire only when it rained. This seemed very odd and we often debated this when discussing our charges. Several theories were promulgated to account for it, but it was sharp-eyed Richard who found the solution.

Jasper completely dominated Mrs Jasper, but when it rained he would retire to the heart of a bush that had grown up in a corner of the aviary. Once he was out of sight Mrs Jasper came into her own. She blossomed temporarily as a personality and she would then display her full artistry as a mimic. Then, and only then, did she feel free to let herself go.

There was, however, an important exception and that was seen after we had moved from Horsley to Albury. There, with a garden ten times the size of the one we had at Horsley, we built a large composite aviary for some of our oldest inhabitants. It was in the form of a huge box, so to speak, of the usual poles and wire-netting, thirty feet by twenty feet high, and divided into four by partitions of wire-netting. In one compartment were Jasper and Mrs Jasper. In the second was Niger, the rook. In the third was a jackdaw. In the fourth a pair of magpies. All mimicked human speech to a varying degree. The jackdaw was limited to its own name, Jacko. The cock magpie would say his name, Oscar, and whistle a tune. His hen would say 'Come on' and give a hearty human laugh. These three also used their native calls. Niger, the rook, had never used his native cawing but always made a curious raucous sound which can only be rendered in print as *wha-ha*, which seemed to be the best he could manage in trying to say Niger. When making this call the whole action of his body was that of any other rook cawing, so *wha-ha* was a complete substitute for the natural caw. He would also say 'Come on'. Jasper and Mrs Jasper had extensive repertoires, as we have seen.

A year after the birds had been established in the multiple aviary we had some thirty visitors to the garden, one Sunday afternoon. Soon the visitors were crowded round the aviary. One or two of the birds in it had 'spoken' to them, to the visitors' amusement. Soon the human visitors were excitedly chattering among themselves, about the idiosyncrasies of the birds. As their babel built up, so did a babel among the birds, until this part of the garden was alive with sounds, rising like an invisible cloud to the skies, a jumble of words and laughter interspersed with Jacko, wha-ha, Come on, and much more besides. People and birds, in the mounting excitement, had shed their inhibitions and even Mrs Jasper, although it was a fine day, was running through the whole of her extensive repertoire. Niger was using far more than *wha-ha* and Come on, which set me on the way to another inquiry.

The truth was, as I established, that Niger had an extensive repertoire but used it only when he was on his own. By hiding in a shrubbery near the aviary, later, I heard him talking to himself using two voices, one a deep masculine voice, the other a higher feminine voice. Although no words could be heard distinctly, he was clearly repeating conversations between Mr and Mrs Ivor Noël-Hume, our friends who had reared Niger from a fledgling and had had to part with him years earlier. Yet the memory of their voices still remained with him.

In other words, among talking birds, some are exhibitionists, like Jasper, others are introverts, like Mrs Jasper and Niger. I had confirmation of this at a later date when I was working one day at the other end of the garden from the multiple aviary. There we had an even larger aviary containing crows, rooks and jackdaws. One of the rooks was Corbie.

Of the two crows in that aviary one was never heard to use vocal mimicry. The other, also a hen, never made a noise like a crow. Instead, she would go through the actions of a crow cawing while at the same time giving forth a long-drawn-out scream. This can be related to the fact that the only sounds she made were of children playing. She would laugh, cry, scream, chatter joyously or petulantly, giving the impression of a group of children at play. In fact, people passing the garden, on the other side of the high wall, used to think children were playing in the garden – but all the noise came from this one crow. Moreover, on some days the 'children' would be playing happily, on other days they would be quarrelling, and so on, so perfect were the imitations.

While gardening, hidden behind a yew hedge, on the day in question, I also heard Corbie. So far as we knew he had no repertoire of mimicked sounds, apart from the clucking of a domestic hen. Now he was holding long conversations and I was able to confirm by subsequent observations that he often did this but only when he was not being overlooked.

In addition to his pleasing personality and his powers of mimicry, Jasper has another claim to fame: his anting. This came as a complete surprise to us for we had several times put ants into his aviary and he had taken no notice of them, except to eat a few. Then, one day, the artist Neave Parker, who was also a keen photographer, came to see us especially to get pictures of Corbie anting with lighted matches and burning cigarettes.

Neave Parker called at the house and I led him to Corbie's aviary. On the way we passed Jasper's aviary. I was smoking a cigarette at that moment and, with the words 'I don't suppose it's any good trying you', I held the smouldering end of the cigarette through the netting of Jasper's aviary. The jay took it, held it in his beak, hopped to the ground and played about with it in the way he plays with any inedible object. He dropped it, picked it up, hopped onto a perch with it, put it down, picked it up again, dropped it on the ground then jumped to the ground and picked up the still smouldering cigarette once again. Then it happened.

Jasper brought his wings forward to their full extent, the wings curving inwards in front of the body, their tips almost meeting in front. The magnificent colouring of his wings, with their blue-and-black checkered feathers, was shown to the full and his whole pose was statuesque. The suddenness of the action was dramatic, the bird's posture was theatrical, and the whole scene was spectacular in the extreme. I could feel, rather than see, that but for his stolid character and his portly frame Neave Parker would have hopped with delight. I felt almost too excited to move, my whole attention on the jay so as not to miss even the slightest detail of the performance. The atmosphere was electric.

By now, Jasper had dropped the cigarette but the display continued. He still held the posture as he hopped forward, then turned and hopped back again, still with his wings held out in front of him, like a canopy shielding the front of his body. He hopped onto a low branch, resumed his normal posture for a moment, then he brought his wings curving forwards again, dropped to the ground and for the next few moments went intermittently into the full display.

So far as I know, this was the first time anyone had seen the European jay anting. Certainly, if we could get a photograph of it, this would be the first time it had been photographed: and the anting posture of the jay is not only spectacular, it is also unlike that of other birds; and the anting display is carried out without an ant in the beak.

How long this first display lasted it is impossible to say. Probably it was less than half a minute, but it had caught us unprepared. We had tried Jasper repeatedly with ants since he was a fledgling. I had tried him with lighted cigarettes and smoke to see if he would emulate Corbie. All had been without result so we had not expected any reaction now and no attempt had been made to have the camera ready for instant action. Moreover, the manner in which Jasper had taken the cigarette, and for a while had merely played with it, had put us completely off our guard. Indeed, Jane and Neave Parker were already on their way to Corbie's aviary when it all started, and had turned hurriedly back on hearing my urgent call.

Frantically, Neave Parker started to take his camera out of its leather case, fit the necessary short-focus lens and bring the camera into action. I was dimly aware of his puffing and blowing and swearing mildly under his breath, as we all stood watching the jay, fearful lest he would tire of this activity, and perform no more, before the desired pictures could be taken.

There need have been no anxiety. To begin with Jasper, unlike Corbie, was not camera-shy. He was an exhibitionist to his wing-tips. I went into the aviary, followed by Neave Parker, and for the next three hours lit one cigarette after another to hand to Jasper, who seemed insatiable. He anted with each cigarette. I smoked myself dry just with puffing enough at each cigarette to get it smouldering well, kneeling before Jasper to do this, while Neave Parker crouched nearby taking shot after shot of the jay.

At the end of three hours, Parker and I returned to the house and sank into armchairs, exhausted but elated. Of the many exciting moments I have experienced since we first thought of making a wildlife sanctury, this afternoon was without doubt the most memorable.

In the days and weeks that followed I tested Jasper with ants, cigarettes and smoke. Once he had performed, on the red-letter day, he continued to ant to order, with ants, with smoke, with cigarettes. It was not the taste of the tobacco that acted as a stimulus. He would not ant with an unlighted cigarette, although he would always grow excited at the sight of one. He would then take it and, using beak and toes, tear it to pieces. Indeed, a cigarette, lighted or not, acted as a magnet to him. He anted most vigorously with a cigarette that had been extinguished but was still warm. He became interested in my hip pocket, in which I kept my cigarettes. In addition, he lost all his aggressiveness towards me. He would once again jump onto my shoulder, tweak my ear, climb onto my head and pull my hair; and he had never before tweaked either my ear or my hair. The whole of his behaviour suggested an ecstasy and the compelling influence of acute sensual or sensory craving. I feel these words are not too strong to describe the bird's moods and actions.

Jasper was with us for seven years before he finally managed to escape from his aviary (we never could be sure how he managed this) and flew southwards, from tree to tree, whistling like the milkman, calling out 'Hello, Jasper'. The sounds grew fainter as he put more and more distance between us, leaving Mrs Jasper a grass widow. My heart sank at the loss of such an entertaining bird and I only hoped he would find romance at last with a second Mrs Jasper. But I doubt it. Having never seen his parents, because his eyes were still closed when he was brought to us, and having been befriended by human beings, he probably regarded himself as a person, not as a jay – and this may well explain his coldness towards Mrs Jasper.

10 *Mainly about rats*

'Can I come in and see your animals?'

It is surprising how often this request has been voiced in the twenty years since we came to live at Weston House – surprising because it reveals how widespread all manner of people's interest is in live animals.

We used to have visitors at Horsley but they were naturalists and practising zoologists. Our near neighbours, or anybody else living in the village could, usually unobtrusively it can be said, satisfy their curiosity by peering over the gate or peeping through the tall hedge that surrounded the small garden there. The larger garden at Weston House is screened from the public highway by a tall wall, so to see our animals now visitors must enter by the heavy seventeenth-century iron gates and walk past the front door by a path leading to a small white gate. Looking over that small wooden gate, the visitor is confronted by a vista of wide lawns, shrubberies and clumps of tall trees. The cages and aviaries are largely hidden from view, so to see their inmates the visitor must be taken on a 'grand tour', which usually occupies an hour or more.

The present visitor proved to be a retired solicitor whose forbears had lived in Albury for centuries, a man I had met briefly a few weeks before at an informal party.

'Do come in.'

As usual the first thing I showed him, just inside the white gate and partly hidden by tall rose bushes and a climbing shrub, was a large rectangular cage, ten feet tall, containing grey squirrels, the charming but pestiferous North American rodent introduced into Britain at the close of the last century.

I was enlarging on the history of this introduction when my visitor noticed an aluminium bowl inside the cage, suspended from the wire-netting a few feet above the ground. It was there to hold drinking-water for the squirrels and, some time before, one of them had started to nibble round the rim of the bowl. The habit had caught on and now a large chunk of the side of the bowl had disappeared; the bowl's capacity to hold water was down to a third of what it had been originally.

This seemed to hold my visitor's attention more than the animals in the cage. I told him I would show him more evidence of metal-chewing as we went round the garden. One of the features of a well-established garden, like this 300-year-old one at Weston House, is the lead labels bearing the names of trees and plants. In these days of plastics and the high price of lead, lead labels are disappearing – not least because they are chewed to fragments by grey squirrels.

As we turned from the squirrels' cage to wander down the path along the north side of the garden, my visitor said he had recently read in his newspaper about rats chewing lead water pipes, and found it hard to believe. I replied, with an unjustified air of authority, that a fat book could be written on all the things rats and mice have been known to bite through.

'After all,' I explained, probably needlessly, 'squirrels, like rats and mice, are rodents, a word meaning "gnawing animals". There are, of course, many instances known of rats gnawing through stout lead pipes.' (These 'grand tours' round the garden of Weston House tend to develop into lecture tours, partly because of my innate garrulousness.)

'Even mice have been known to do so,' I continued, warming to the subject. 'Electric cables have been similarly treated. The lights suddenly go out in a house (even when there is no power cut!), the electrician is called and ultimately, after half the house has been pulled apart in the search for the fault, a cable is found bitten through and beside it is the body of a mouse, electrocuted. Then there are the familiar holes bitten, by rats more especially, through stout wood, plaster, even concrete.'

By this time we had reached a cluster of small aviaries under the spreading branches of a copper beech. This was our first-aid post where animal waifs and strays, brought to our front door, often by children, were placed as the first stage in hospitalization. At that moment they contained only a few common or garden birds in need of attention. My visitor gave them no more than a cursory glance, so that the lecture on rodents flowed almost uninterrupted as we made our way along the remaining hundred yards of the path to the particularly large corvid aviary housing our crows and a rook.

I remember the first time I had become actively aware that rats would chew through concrete. It was soon after we had started to build up the menagerie in our garden at Horsley. We needed a shed in which to store the various pieces of equipment we were beginning to accumulate, so I had bought a pre-fabricated one, and a concrete foundation was laid on which to site it. Within a short space of time rats were using the shed. They had not eaten their way through the wooden walls of the shed, as one might have expected, but had taken the more difficult route and tunnelled through the concrete.

Just as rats infest a farmyard, taking advantage of the food lying about, so they had come into our garden, for inevitably there are scraps and items of food scattered around when one has animals needing to be fed. There are also other pickings to be had – the food of the three melanistic pheasants we had been given, for example.

In the north-eastern corner of the garden was a dry stone wall and, for want of space elsewhere, we had erected an aviary for the pheasants in front of it and included the wall as part of it. Pheasants require grain as part of their diet – and, inevitably, the rats took up residence within the wall, making their exit holes all over the face of it. If you watched carefully you would see a rat's face appear at one of these holes. It would stay there for several minutes, provided you kept very quiet and absolutely still, then three or four other rats' faces would appear at other holes in the wall. They would also be still for a while. Then other faces would appear at other holes. Finally, as if at a signal that all was well, the rats would emerge, climb down the wall and start to feed.

I had become interested in the long-standing story that two rats can combine to carry away an egg: one rat lies on its back, clutching an egg with its four legs, and is dragged along by the other rat. So many people have written to me or told me they have actually seen this happen that I felt there must be some foundation to the story.

Very soon the two hen pheasants of our trio laid two dozen eggs in the same nest on the top of the dry stone wall. The wall was nearly five feet high, yet these eggs disappeared one by one leaving no trace, no splash of yolk, no fragment of shell, no broken eggs on the ground below. It was as though they had evaporated into thin air.

I discussed this with Jane. There was a definite path across the floor of the aviary where the tramp of tiny feet had stamped down the earth into a typical rat-run. We decided to place an egg beside this run, taking care that the scent from human fingers should not be left on the egg. Then we would take turns sitting, hidden but watchful, near the aviary, with a camera ready. Our minimum hope was that we might get a photograph of a rat in the act of transporting an egg; at that time it was not even known how this was accomplished. Our maximum hope was that we might see a rat lie on its back, clasp the egg to its chest and be dragged along by the tail by a fellow rat. And we might even get a photograph of it!

Watch was kept for a week, throughout the hours of daylight, and I have to confess Jane did most of the watching. Rats were seen passing and re-passing the egg, but it remained stolidly in the position in which we had placed it. At the end of the evening of the seventh day we came to the conclusion that precious time was being wasted. We decided to terminate the exercise.

The following morning the egg was gone. It had vanished without trace.

Although we failed to see two rats in tandem, either then or subsequently, my faith in the well-known story remains tolerably firm if only because of an encounter I had after we had moved to Weston House.

It happened that one visitor, who came specially from several miles away to look round our menagerie, specialized in wildfowl. She invited us to come and see her collection. On our second visit she told me the story of the man who visited her periodically to deal with rats infesting a wooden shed where she kept grain for feeding her birds. Having heard what she had to say I asked if it were possible to meet him, to get his story first hand.

A few weeks later she telephoned to say he would be calling on her the next day. I was introduced to him casually and after a few banal remarks I said: 'I hear you had a surprise encounter with a rat.'

'Yes,' he said. 'I opened the door of this shed and as I went in I saw a rat run across the floor and disappear behind a pile of sacks. At the same time, out of the corner of my eye, I saw a piece of sacking on the floor near my foot with an egg lying on it. The next moment the sacking turned over and ran away. It was another rat.'

It is hard to believe anyone would have the ingenuity to make up such a story, much less tell it to someone if he did.

By the time I had recounted this episode to my visitor we had reached the far end of the garden where there was our largest aviary, a really huge one, containing two crows and a rook. Adjacent to it was the pen containing our pair of foxes. We duly inspected these and talked about them before setting off for the south-east corner of the garden where there was another large aviary containing bantam chickens. It was on the opposite side of the kitchen garden from the foxes' pen.

'How do they manage to chew through concrete?' my visitor asked, clearly reverting to rats. I tried to explain.

Human prisoners, in novels at least, have gained freedom by scraping away the mortar with a stout nail, thus loosening the bricks of their prison walls. All that is needed is endless time, persistence and a sharp instrument. One of these, at least, every rodent has in full measure. Moreover, the rodent has the advantage that the sharp instrument it uses does not wear out, for it consists of a set of incisor teeth, two in the upper jaw and two in the lower. The usual teeth have an inner soft dentine covered with hard enamel. The rodent's incisor has enamel on the front only and the dentine wears away more quickly than the edge of the enamel, giving a permanent chisel edge. Then again, the incisors are continually growing at their roots, and unless used constantly will not keep in working trim. Whenever a rat is not feeding or doing something active that requires all its attention it is gritting its incisors. There can be little doubt this is a rat's way of keeping its teeth sharp – and it may be the characteristic behaviour of all rodents.

The mention of the incessant gnawing of rodents recalled an occasion at Weston House, when we had been there two years. There is arable land on two sides of the garden; that year it had been put down to wheat and the wheat had been harvested. The morning after the last wheat had been cut the kitchen garden, which occupies about a quarter of the whole garden, was infested by rats. Presumably, as the combine harvester had gone round and round the field, working towards the centre, the rats had gone to earth. Then, during the night, we must assume, finding their habitat devastated and denuded of food, they had set forth to look for fresh supplies and they had found them. I tried to describe the scene to my visitor.

'There were rats everywhere in the kitchen garden, scores, possibly hundreds, of them. They were gnawing mainly at the root crops but, most remarkable of all, they were climbing to the tops of the Brussels sprouts and steadily eating them down to stumps. If our view was correct that they had been living in the wheat field, most if not all of them had never been disturbed by people so they did not recognize a human being as a potential enemy. Consequently, as we walked among them they did not bother even to get out of our way.

'Above all, and this is probably my most vivid recollection, the air was filled with the sound of their gnawing. The sight of the rats was bad enough. The sound of their gnawing was nauseating.'

It was a great regret that Jane was not on hand to see this. She had recently married and was at that time in West Africa. Ironically, her husband is a specialist on rodent control and had been sent out to advise the Nigerian government on this very subject. Jane was with him, so were her cameras and so were her enterprise and energy. I have often reflected on what photographs she would have taken and what use she would have made of this golden opportunity, when one considers what she did with a lone rat that wandered into one of our aviaries.

It was a full-grown male, quite formidable to look at, and Jane was determined to photograph it. But to do so we needed to transfer it to an aviary with smaller mesh netting, so that it could not escape. While I stood pondering the problem Jane had lifted the lid from a grain bin standing beside the aviary, stooped down, grabbed the rat by the tail and dropped it into the bin, from which it could not escape. The whole manoeuvre had taken little more than the blink of an eyelid. Thereafter Jane spent many hours in the aviary photographing the rat, once it had settled down in its new quarters. Had she been in the garden surrounded by rats that did not bother to move away she would have been in her seventh heaven.

Unfortunately they left us suddenly, when none of us was on hand to witness their departure. However, a neighbour, who happened to be walking past at the time, saw what she described to me later as a mass of rats issuing from our garden. We also heard from the scientists at the research laboratory – the ancient mill two hundred yards down the road from Weston House, that has been converted into a commercial testing laboratory – that 'an army of rats' had been seen beside the River Tillingbourne that runs through the laboratory grounds.

On three separate occasions during the previous twelve years, somebody had written to me about seeing a phalanx or a column of rats on the move; I had always hoped I might see this phenomenon for myself. Now we had had such an event on our doorstep and I had failed to witness it!

At about the time that the rats had invaded our garden there appeared in the *Veterinary Record* (8 July 1961) a letter written by W. J. Tarver. He gave a few more accounts of columns of rats on the move. His informants were farmers, and one of them had spoken of a chirruping and whistling noise made by the rats as they marched along. A second spoke of going with two friends and a couple of lurchers, one moonlit night, to hunt rats infesting a poultry farm. In the farmyard they found unusual activity among the rats. They turned to set the dogs on them but these had disappeared. Then they heard this same strange chirruping and whistling coming from the rats: it grew gradually louder and eventually the head of a column of rats advanced into the farmyard. Now the noise reached a crescendo and, as the column moved across the yard, scores of rats tumbled out of the surrounding buildings to join the column, apparently attracted by the sound. The column took twenty minutes to pass the three men. Some of the rats ran between their feet; they stood stock still, scared but riveted by the unusual sight. No rats were seen on the farm for the next six months.

Another farmer, in another magazine, described the sounds made by a column of rats on the move as 'a noise like a thousand starlings chattering in a tree'.

Mr Tarver continued: 'It is not conceivable that a man with a highly perceptive ear and a knowledge of wind instruments could construct a pipe which would duplicate the noise of a rat migration?' He also suggested that a tape recording of it might serve to lure large rat populations to their doom. Although he did not mention it, Mr Tarver's thoughts must have turned, as mine did on reading his letter, to the Pied Piper of Hamelin. This fanciful poem could well have been founded on more solid fact than we normally suppose possible.

My visitor cleared his throat, bringing me back from my silent recollections. 'Tell me more about rats' teeth,' he said, so I relinquished my exploration of memory's pathways and resumed my impromptu lecture.

Whether we take the view that it was Providence or the random mutation of a gene that gave the ancestral rodent its chisel-incisors, there can be no argument that these teeth led to unparalleled success. A rodent can chew anything edible, soft or hard; it can open nuts that defeat other animals, for example. It can burrow into hard ground or other hard substances. The result has been that rodents have flourished, producing more species than any other group of mammals, and populations, which in some species, such as brown rats, exceed even the vast human populations now causing severe headaches to economists, politicians and others.

The same agency, whether Providence or a gene, or one of these in combination with natural selection, not only gave rodents efficient incisors but the urge to use them constantly. In a world untouched by human hands, this is a virtue. It not only ensured the rodents' survival by opening up to them limitless supplies of food and enabling them to burrow into anything but hard rock, but it also fitted the rodents to be first-class scavengers. This comment is based on the assumption that rodents gnaw more persistently than is necessary to obtain sustenance or shelter. There are no statistical studies to support this assumption, but one has the general impression that much of the damage to man's crops, buildings and other possessions falls under the heading of wanton damage. That is to say, damage not wholly connected with the rodent's necessities for living, unless keeping the incisors sharp and wearing them away to counteract the growth at the roots be counted as such a necessity.

My reflections were cut short by another remark from my companion: 'If they chew lead pipes do they suffer from lead poisoning?' This turned the spotlight on yet one more advantage held by rodents, that they can chew incessantly and indiscriminately without necessarily paying a penalty in gastric disturbances. Most animals have teeth forming continuous rows from the front of the jaw to the back of the mouth. In a rodent's dentition there is a wide gap of toothless gum between the incisors and the cheek-teeth. The lips can be drawn into this gap, on either side of the mouth, completely shutting off the front of the mouth from the rest. When gnawing lead piping or other harmful substances, the chips are shed from behind the incisors out of the mouth.

Much of what has been said here may be familiar, and yet it is apt to elude one at an important moment, and this has led to rather unusual consequences. Rats are given to gnawing bones, even old bones from which every vestige of flesh has disappeared. It is part of the natural scavenging service already mentioned. It may be that the rats obtain materials essential to their bodies; or it may be merely due to the rodents' urge to gnaw. Archaeologists have found, on occasion, among flint tools and fragments of pottery, bones that bear the appearance of having been worked by human hands: they seem to have been specially shaped and the markings upon them might have been caused by chipping with flint tools. Archaeologists are no less cautious than other scientists and seek confirmation of their ideas from other sources. And when such bones are shown to a zoologist it is not long before he realizes that the bones have been gnawed by rats and buried, by chance, on the site of an early culture.

Why rats should gnaw bones has not been subject to close scientific scrutiny, but it is a fair guess that, as with so many other mammals, calcium is needed by the body, and that bones offer a ready source for this. It may be, on the other hand, that the gnawing of bones is, like gnawing through lead pipes, a non-utilitarian pursuit, a result of hyper-activity on the part of the rodent that has no functional end. In 1954, two American scientists set to work to assess the differences between the domesticated rat and the wild Norway or brown rat. We know that the first is gentle and trusting, does not bite unless frightened or hurt and makes no attempt to escape. The wild rat, by contrast, is fierce, aggressive and suspicious, attacking on the slightest provocation. In captivity, it will take the first opportunity to escape and always remains suspicious and tense. These marked differences in behaviour are the result, among other things, of a considerable difference in the adrenals, those of the domesticated rat being only one-third to one-fifth the size of the adrenals in the wild rat.

Another difference the two scientists found was that when deprived of food the general activity of the domesticated rat increases by no more than 32 per cent, whereas that of the wild rat increases by 142 per cent; that is, it becomes four times more active than the normal. The control rats, of both domesticated and wild strains, that were given adequate food showed the same amount of activity, so these authors concluded that 'in both the domesticated and the wild rat, the absence of food would tend to increase the activity because of the more active adrenals, but owing to the greater size and activity of the adrenals in wild rats, their increase in activity would be greater'. From this, it is not difficult to visualize a hungry rat gnawing the first thing it comes across, whether it be plaster, concrete, lead pipe or bone. In some instances the gnawing may be purposive, to tunnel in an endeavour to search for food. In others, it may be merely due to the impulse to gnaw resulting from an empty or half-empty stomach.

The truth is that I was carried away by what has been called the exuberance of my own verbosity. My temporary companion indicated this by breaking in with 'All very interesting, but I can't stand rats'. So I told him about the only occasion when I had seen a brown rat living what must have been its natural life.

We are apt to think of rats as inhabitants of towns and barns. This rat sat in the crotch at the base of an alder tree. It was leading a simple life in the heart of wild fields with its bolthole in the bank of a quiet stream that flowed just below where it was sitting. It was not easy to see the rat for the long grasses with their two-foot stems, their heads heavy with seeds overhanging the water. Then the rat sat up on its haunches and, using its front paws, took one of the stems and hauled it in, as a boatman would haul a mooring rope. Then, holding the stem just below the seed-head, it nibbled from one end of the seedhead to the other until not a seed was left. After that it hauled in another stem and ate the seeds in the same way. I squatted on my hunkers and watched, fascinated, as it treated one after another of these stems in the same way.

Rats are naturally seed-eaters and this one was showing me how its ancestors subsisted, before mankind started to grow and store cereals, so setting the species of *Rattus* on their long path down the ages to become parasites, pillagers and destroyers of grain and eaters of garbage and carrion. This rat was clean, its underbelly as it stretched up to haul in yet another seedhead showed pure white. Its paws were clean, a whitish-grey and remarkable in the way they were being used as hands. Here was the rat, the pest, in its natural setting – a clean, harmless and quite charming wild animal.

'I hate rats,' broke in my new-found friend once more. 'I find them repugnant, loathsome.'

Well, he started the conversation about them. Yet I could only agree with him and I searched my mind to see why this should be. After all, there must be few of us that have not, at some time in our youth, kept the domesticated form, the white rat, as pets or have handled those kept as pets by someone else.

Then I recalled the incident that probably laid the first seeds of repugnance in me. One of my earliest recollections, as a small child, was hearing my parents discussing an elderly woman who had died alone in her house. Her body was discovered some days later, and rats had gnawed the flesh from her face and hands. Distasteful this may be but it points to a natural function of rodents: seed-eaters primarily and scavengers as a subsidiary. We by our wasteful habits, our unclean ways, have brought out the worst in one of Nature's scavengers. My perfectly charming rat in the alder crotch had done quite a lot to offset in me the worst aspect of our world-wide rodent pest.

Another feature of rat behaviour also helps to put the animal in correct perspective. This is what one writer has called their diabolical cleverness. We saw a good example of this one year at Weston House, and I felt impelled to tell my guest, especially as we were at that moment near the spot where it happened.

From time to time we would see from the house a black-and-white cat, belonging to a neighbour, making its way down the garden. Wondering whether it had designs on some of our aviary birds we investigated. The trail of the cat led us to the yew hedge.

Rats were coming in from the adjoining fields to scavenge the compost heap, which stood on the south side of the kitchen garden, at the end of the yew hedge. They had a regular run along the base of the hedge, for a distance of thirty yards. Now, instead of running along the ground they were making the journey through the top of the hedge, eight feet up from the ground, moving as fast and as skilfully as squirrels. Halfway along the hedge is a gap and over this droops the lowermost branches of a tall Wellingtonia. As each rat came to this gap it leapt from the yew, up to the Wellingtonia branches, and then dropped down to the next run of the hedge.

And what was the reason for this energetic, acrobatic display? The cat was crouched, day after day, with its nose near the rat run at the base of the hedge, patiently waiting. In the end it gave up this fruitless pastime.

Some time later we found that a large rat was taking potatoes and apples stored in a brick-built potting-shed halfway down the garden. We could see the rat's trail across the floor and out through a hole where a brick was missing. We tracked its runway from there along the base of a paling fence and into a derelict wooden shed just beyond our boundary – and there we found the missing objects, some nearly whole, others partially chewed. How had the rat managed to transport quite large potatoes the seventy yards from our brick shed to this derelict shed?

I consulted the literature on rats without success, so I started to inquire around. Nobody seemed to know. Eventually, however, I managed to find people who had actually seen rats in the act. It seems that they use several methods. They may use paws or snout, or both together to bowl a large rounded object along the ground, or they may hold it between the paws, with the chin assisting, walking on their hindlegs with the tail pressed on the ground giving the body additional support. Since I was never able to catch our rat in the act I spread a thin layer of nearly white, fine sand over the floor of the shed in which the apples and potatoes were stored. The tracks were examined each morning and the sand smoothed over again. These tracks showed my rat to be using all these methods at one time or another. At one point along the trail leading to the derelict wooden shed the rat was obliged to pass under an unsurmountable obstacle. It had tunnelled underneath it and from the diameter of the tunnel it seems safe to presume the rat must have pushed its loot through with its snout.

One of the many people of whom I inquired, gave me a detailed account of watching a rat transport an egg a distance of twelve feet. The rat used its body as a conveyor belt. It lay on its back, picked up the egg between its hindlegs, pushed it along its belly, transferred its grip to its forelegs and pushed the egg beyond the level of its snout. Then it got to its feet, moved forward and repeated the manoeuvre. I saw no sign in the sand that my rat had thought of this.

By this time we had arrived near the end of our grand tour, in that part of the garden where, among the shrubberies, were most of our aviaries. There were owls, magpies, jays and others – at least they took our attention off rats for a while. Finally we were back at the little white gate. The tour was at an end.

Anyone who persuades me to guide him or her round our garden must suffer the possibility of being surfeited with my tales, stories and theories. On this occasion something my visitor said, as he was about to take his leave, launched me on what was at that time one of my minor hobby-horses.

In 1917 I was in the line that ran in front of the town of Béthune, in northern France. Rats infested the trench and our dug-out. My companion developed an almost obsessive enmity towards them and when off-duty he used to hunt them with a revolver. One day he told me how he had cornered what he described as a big, old male rat in the angle of a trench. As he raised his revolver to shoot, the rat backed into the angle of the trench, 'its eyes almost popping out of its head, and screamed. It was so like a baby crying,' he went on, 'that I could not pull the trigger and I lowered the revolver and let him escape.' He described the noise as piteous in the extreme, a cry that tore at the heart-strings and, recalling this episode, I am wondering whether this is the answer to the conundrum I raised in chapter eight about the screaming of a trapped hedgehog. Perhaps it is not so much a cry for help as an involuntary cry in order to deflect further attack. If such a cry can call forth pity in a human being bent on ruthless destruction, there is just the possibility that this same cry might have a similar effect on an animal predator. The point is worth pursuing.

We are told that a frog pursued or seized by a predator will utter an unearthly scream. What we are never told is whether that predator calls off its attack or releases its victim. Similarly, we are told that a rabbit pursued or seized by a stoat will give out a heart-rending scream. I have never heard either of these. What surprises me is that a rabbit, normally fairly silent, at most giving out low sounds which are almost inaudible except at close quarters, should scream then. The European common frog, for example, does little more than croak and that only in the breeding season. I have, however, heard what may loosely be called cries of anguish from other animals in distress. Are they 'cries of anguish' or anything approaching this? The answer is that they might be cries of aggression or something else but that nobody knows for certain.

Perhaps our half-Siamese cat provides something of a clue. She is not so fully vocal as a true-blooded Siamese but she does 'talk' a lot and there is nothing remarkable in any of these sounds. They are variations on the ordinary mewing. There is, however, a note of urgency if feeding time is near, and when she is extra hungry this note of urgency becomes even more emphatic. She will seek me out and try to attract my attention by standing in front of me and looking up at me. She will then start to walk in the direction of her food bowl, stop, look round at me and, if I show no signs of following her, return and repeat the manoeuvre. Should I continue to ignore her, she claws at my trouser leg as if to drag me in the direction of her empty food bowl. Throughout these actions she keeps up an almost incessant mewing, with the notes of the mewing assuming a tone of increasing urgency. Then comes a cry, almost a cry of anguish, as if she were saying: 'Oh, do please hurry. I am so hungry.'

What has struck me forcibly is that this cry of anguish, as I have called it, is almost identical with a phrase of music, namely, the most poignant phrase, which is several times repeated, in Dido's lament, from Purcell's *Dido and Aeneas*, in which Dido is expressing her grief. As soon as one becomes aware of this similarity and runs over, in memory, passages in laments written by other composers, and then recalls the 'cries of anguish' one has heard from trapped animals, the comparison is most striking.

My ex-solicitor friend appeared thoughtful but made no comment. Finally, he thanked me, shook hands and said: 'Goodbye.' The catch on the little white gate clicked behind him.

11 *Duck tales*

Some time ago I received a letter from Florence Lippiatt of London, who described what she had seen while standing at the Buckingham Palace end of the lake in St James's Park:

> I noticed a duck with her brood; they were, or so it appeared, just out of the nest. Nearby were other ducks and one was diving. One of the ducklings watched it, you could almost see the surprise written on its face, then it tried. It was comic to watch for it couldn't manage it. So it watched again and this time, which was about its seventh try, it was successful. Was it pleased? Tremendously so. Then it looked for its parent and the rest of the family. They were nowhere to be seen, for they were making their way to the other end of the lake. 'Full steam ahead', we could hardly keep pace. By the time it had found the family it was exhausted. There was no more pride in what it had achieved and it just meekly followed in the wake of the others.

After I had read this I felt that somebody must have had an adventure of this sort for the idea of Donald Duck to have been conceived. It has always surprised me how popular Donald Duck has been, probably the most popular of animals since the cinema and later television became familiar in the lives of people. Donald Duck is a white domestic duck but his forbears would have been the wild duck or mallard so widespread throughout the Northern Hemisphere. I have known a number of people who have befriended mallard or have kept them as pets and I can recall only one instance in which a duck showed any unusual personality.

It was many years ago that I heard of this mallard drake, inevitably named Donald, that became almost a household pet. According to what I was told it followed its owner's two dogs wherever they went, it shared their food, it made a habit of sleeping on the back of one of them and when the dogs went into the water to swim it would ride on their backs holding on to the hair on the top of the dog's head with its beak, changing if necessary from one dog to the other when the first dog had had enough of the water and climbed out onto the bank.

Mallard are wild but they readily settle down on water near

human habitations and can easily be tamed by regular feeding. For many of us they provide our first introduction to natural history, one might almost say to wildlife, except that ducks are so tame. Thus, it is a common thing for children, almost as soon as they can toddle, to be taken to 'feed the ducks', in parks or other places. In fact, juveniles and adults alike derive a great deal of pleasure from throwing pieces of stale bread to the ducks.

There was an unhappy instance to do with ducks that occurred after we had been at Albury for several years. The Tillingbourne River runs through a pheasant preserve there, and then on through the village where a number of gardens on the north side of the river extend down to its bank. Where the Tillingbourne runs through the pheasant preserve the head keeper regularly rears a number of mallard. These migrate downstream and the people living in the houses on the north side of the village feed them and make pets of them.

There came a day when a pheasant shoot was being held on a Saturday afternoon. The 'guns' belonged to a syndicate who paid a fairly substantial fee for its members to shoot the pheasants. On this particular Saturday afternoon we heard the banging of the guns all around, a commonplace sound during the pheasant season, and we paid little attention to it. Then, towards the end of the afternoon, we became aware that the mallard that had that day been haunting the village were flying over in the direction of the guns: they had been driven there by the beaters, and something in the nature of a massacre followed.

The people in the village whose children had been feeding the mallard were furious. I spoke to the head gamekeeper subsequently and he confessed that he had no idea that the ducks were being petted and tamed, that the members of the syndicate on that afternoon were such poor shots that their bag of pheasants was negligible and that he had to send them home with at least something to show for their afternoon's shooting. So he sent his beaters down to round up the ducks.

Soon after we had first had the idea of a sanctuary, Jane suggested we might keep some mallard. This was slightly contrary to our original intention, which was to encourage wild species of animals into our sanctuary, and later to use our resources for rescuing birds and other animals that were injured or in other ways needing help. What was in Jane's mind, however, was a deeper plot. She was taking film shots of the various animals we had in the garden and she thought it would be a good idea to have at least one shot of an aquatic bird. I got in touch with the then Secretary of the Parks Committee of the London area and a few days later he telephoned me to say that if I went to Kensington Gardens, to the Curator of Wildlife there, he would be able to let me have a pair of mallard.

I duly made the journey and the Curator told me that he had more mallard on the Ponds than he needed – and he mentioned the reason why. One of the residents of nearby Bond Street, living in a fifth-floor flat, had got in touch with him because a mallard duck had laid a clutch of eggs in a large empty flowerpot standing on one of his windowsills overlooking the busy street below. He was anxious to know what he should do about it. The Curator's reply was brief and practical.

'Wait,' he said, 'until you see the eggs are hatching. Then put a cloth or a piece of cardboard over the top of the flowerpot and bring the whole thing to me. If you don't, when all the ducklings have hatched out the duck will fly down into the street and call to them. They will then drop down one by one from the windowsill and the duck will lead all the ducklings to the nearest water. In your case, this will be Kensington Gardens. Either some or all of them will be run over, or else there will be a hold-up in the traffic to let them pass, and it is far better to anticipate this and apply a remedy, which is to bring them to me.'

Anyway, I selected a pair of mallard, picked them up and placed them in a large hamper basket which I had brought for the purpose. I looked forward to the journey home with them with some misgivings. What would happen, I thought, if the ducks start quacking when I am on the station and a railway official hears it – am I going to be charged the fare appropriate to live-stock travelling by train? Likewise, will we have a chorus of quacks in the train itself during the three-quarter-hour journey so that I have to explain to all and sundry in the railway carriage why I am transporting live ducks in a basket?

In the event, the ducks were silent, except for the occasional subdued quack from the basket on the luggage rack. The only reaction of my fellow passengers was for one or other of them to cast a furtive eye at the basket and then bury himself again in his newspaper.

The two mallard were placed in a fairly roomy aviary pending the construction of a pond. (We put them in an aviary rather than pinion them.) The duck-pond was duly made and our plan was to build round it a large wire enclosure; but the plan, as I have mentioned earlier, misfired because the two men I employed to cement the pond were less skilled than they made themselves out to be. The pond never held water, for as the cement dried it cracked in all directions. Meanwhile, the mallard were provided with the best substitute we could find – an iron bath, oval in shape, three feet long by two feet across and a little over a foot deep. A ramp was arranged on one side to enable them to enter the water in comfort.

During March and April of the following year they performed some strange evolutions in this limited stretch of water. Both duck and drake would enter it together and swim round and round each other, bobbing their heads up and down. A film shot of this was included in one of a set of films that I used for lecture purposes for many years. As soon as the two mallard were seen on the screen swimming round and round each other in so limited a stretch of water, the audience, whether juvenile or adult, would burst into roars of laughter, and I always waited for the laughter to die down before I made the comment: 'The Swiss Navy'.

Years ago it was not infrequent for somebody to refer in a ribald way to the Swiss Navy, to epitomize something that was non-existent. The assumption was that landlocked Switzerland could not have nor need a navy. Yet I understand that they do have patrol boats on their lakes where these extend into the territory of neighbouring nations. My comment therefore was fairly appropriate because these patrol boats must spend much of their time going round and round in a fairly limited space.

Two years later, when we moved to Weston House, we were able to build a real duck pond. We had five mallard by then. They were penned up at night, and each morning they were let out again. Their first action then was to go to the pond and bathe. On one occasion they were let out much later than usual, through an oversight. They all ran down to the pond and in no time at all they were bathing so vigorously that the surface of the pond was in a turmoil, a seething cauldron with five ducks almost obscured by the water thrown up. The time was mid-winter.

It so happened that my wife and I were standing at the window while this was going on, and after watching it for a while she said: 'You can't say those ducks are not enjoying it.' This incident sprang to mind some time later after I had given a lecture. At the end of it a young man, an undergraduate in biology, came to me and asked: 'Would you say that animals ever do anything because they "like" doing it?' And then he added, without waiting for me to reply: 'I get tired of this attitude that no animal enjoys doing anything.'

I imagine the student was reacting against what I have heard described in non-scientific quarters as the 'arid language of science'. For the purposes of study it is essential to use terms, in describing animal behaviour, that give the greatest precision of thought. They are also necessary to avoid the teleological approach and anthropomorphism. In other words, you must not attribute to animals, even to the higher animals, human desires and emotions. This is reasonable enough, up to a point, but it contains within it the dangers of over-simplification. The fact is that you can't have your cake and eat it too. If man is one with the animal kingdom, as current biological thought demands, and if there has been continuous evolution of all living things, then clearly the roots of human behaviour must be identical with those of all animals. The tree may have blossomed more magnificently, but that is beside the point.

Bathing is as good a yardstick as any other animal activity against which to try to answer the question whether animals – any animals at all – are capable of enjoying something, or are capable of doing something because they 'like' it. Bathing would seem, at first sight, a strictly functional activity, a mere matter of cleanliness. That is, it is something that must be carried out because it has a value for the survival of the individual – using the word 'survival' in a broad sense. Yet, as I once heard a lecturer on dietetics say, all bodily functions should be enjoyable. When we satisfy hunger, which is the body's call for more fuel to maintain a supply of energy, we ought to enjoy eating. In fact, the aesthetic appreciation of food does more than gratify the senses, it stimulates the processes necessary for digestion – or so we are told.

Many birds, probably the majority of them, are given to bathing in water. That nobody can deny. It is, however, arguable whether they always bathe for the sake of cleanliness or even merely to fulfil some bodily function or need. The subject has been little studied and I have only personal observations to draw upon, but as these cover a period of several years during which I have been interested in the matter they may be adequate. The first thing I have learned, which must be obvious from the start, is that members of some species bathe more than the members of other species. The second thing is that within the same species some individuals are more ready to bathe, and will do so more consistently, than other individuals having the same opportunities.

The next thing that strikes me is that birds belonging to species in which bathing is normal are sometimes kept in captivity and given either no, or very few, opportunities for bathing. Yet these same individuals seem to lose nothing, unless it be an added enjoyment in life – but that is precisely the point I am trying to examine. Certainly they seem to suffer little in health although their plumage may not be in such fine condition.

The observation that strikes me most forcibly, however, is that birds living in temperate latitudes bathe more, and with a greater vigour, on cold and frosty mornings, or when the sky is overcast or there is mist. This does not mean they do not bathe when the weather is hot or sunny but that they are more given to it when the reverse conditions obtain. Our experience with rooks supports this idea. All the tame rooks we have had have been confirmed bathers, especially in cold weather, and under conditions of hard frost we have seen them with icicles on their feathers after a bathe, the icicles rattling as the birds shook their feathers.

Bassie, our grey parrot, would not voluntarily bathe and strongly objected to being sprayed with water. When her tail feathers became dishevelled they would resume their normal appearance on being dipped into water. This presents a paradox: the bird was quite healthy and in good condition yet it was clear that water, or presumably bathing if she would indulge in it, would do her plumage good.

This is not the only paradox. Cormorants spend half their active day perched on the rocks and the other half in the water. Much of the time they spend in water is occupied in diving. When, however, a cormorant bathes it does so in a most thorough-going manner and its bathing session lasts a long time; I have watched one bathing continuously for a full half-hour. It may be argued that such ablutions are necessary to clean the feathers of scales and slime from the fish the bird has caught and eaten. I have no proof whether this is or is not so, but it seems most unlikely from the way the cormorant feeds that it soils its plumage in the process, certainly not enough to merit half an hour's bathing.

Gulls, too, eat fish, in addition to many other things. They spend a great deal of time on the water and they also indulge in desultory bathing at odd moments during the day. But the occasions when I have witnessed the most vigorous and prolonged sessions of bathing have been when flocks of gulls, anything from twenty to fifty strong, come to pools of fresh water forming on the shore at low tide from a landward stream. At such times the gulls bathe and then come on land to preen. They then return for another bout of bathing, follow this with more preening on land, and perhaps alternate these two parts of their toilet several times. All birds preen after bathing, but most do not bathe again immediately after having preened. More important, when gulls alternate bathing and preening there is an unmistakably mounting excitement that grows more intense with each succeeding immersion in the water.

In these and other observations there seems to be evidence that bathing in water can satisfy several needs. It keeps the plumage in good order and doubtless has a cleansing effect too. It would also seem to provide a stimulating effect to the body. This seems likely from the way birds bathe more in cold weather, and from the way gulls will deliberately visit freshwater springs on the shore, which are presumably at a lower temperature than the sea water. There is also the evident air of mounting excitement, not only with gulls but also with ducks.

To enjoy something means to find pleasure in or derive satisfaction from it. To enjoy and to like are practically synonymous terms. Ducks, gulls, cormorants, and the rest, probably bathe both for cleanliness and for enjoyment, just as we do. The only real difference between them and us is that we are conscious of what we are doing and can analyse the sensations we are experiencing. That may give added enjoyment but that is all.

Another marked characteristic of a group of mallard is the noise they make, especially during the night. We are familiar with the fictional Donald Duck, a male, with his loud quacking imitations of human speech, yet it is the female mallard that quacks, the drake being capable only of a low, soft call, as if suffering from laryngitis. This was brought out by the time when, on the Tillingbourne opposite our house, there was a duck and twelve drakes. The noise made by this solitary duck escorted by her dozen drakes was bad enough, but it so happened that on my own pond there were at that time thirteen mallard, several of them ducks, and when they answered the calls of the wild duck the combined noise gave us sleepless nights. We had given up penning them in at night, but now it was decided to shut them in once again. This did not inhibit their quacking, but at least it made it less audible. It also gave us an interesting sidelight on the mentality of ducks.

The mallard had become so used to spending the night in the

open on the pond that when I went to pen them up that first night they had some difficulty in comprehending what was required of them and it took several minutes to shepherd them into the wooden shed where they were to spend the night. The next night, however, as soon as they heard my footsteps along the gravel path there was no difficulty. They just made straight for the shed and went in. The third night they went in of their own accord without waiting for me to appear. So I no longer think of ducks as wholly stupid although they can do many things which, from the human point-of-view, seem the height of stupidity.

After having penned the mallard up several nights in succession, the memory of interrupted sleep became dim. I grew careless, and no longer bothered. So the ducks had their freedom at nights until the next time they gave me a sleepless night. Then would follow the same sequence: the first night panic, the second night the ducks would go into the pen as soon as they heard my footsteps, and the third and succeeding nights they would go in of their own accord. This pattern was repeated several times, always with precisely similar results: failure on the part of the ducks to interpret my wishes, ready acquiescence the second night, anticipation on the third night. It is easy to understand that there is in this something of the Pavlovian conditioned reflex, but there is also mixed with it a capacity for learning readily. There is also a curious antagonism in the matter of memory. My own guess would be that a duck has a short memory: capable of remembering a lesson provided there is a lapse of no more than a few days, but in a week forgetting all it has learned.

In this matter of noise at night, I have the fullest sympathy with someone who wrote to me about a 'lone female duck who resides near a wild colony at the bottom of our garden . . . who starts to serenade us with noisy quacks from dusk to dawn (giving us) a few sleepless nights listening to her.' The rest of the letter, however, was concerned with events that preceded this nuisance.

My correspondent continues: 'Once, she (the lone female) was happy and contented, until the loss of six eggs in her nest, stolen by water rats. Then she began to quarrel with the other feathered creatures and refused to co-exist. Instead, she chases the moorhens and the water rats; the stately swans and cygnets have learned to keep their distance; the noisy seagulls now need no second warning to move farther downstream. "Madam" has installed herself as Queen of the River.'

This duck was showing a reserve of courage of the kind shared by animals and humans alike: as a result of a personal injury or fancied slight we strike out at the nearest living being. It is not the first instance I have met of a duck, having had its eggs stolen by rats, thereafter declaring war on all rats she meets. What is surprising is that this one should have so far gone berserk that she intimidated swans and gulls alike. At least her memory was not short!

This reminds me of the time when Jane and I were staying in Brussels. We had gone there at the invitation of Dr Victor Van Straelen, as guests of the International Union for the Conservation of Nature. We were staying in the Grand Hotel which is next to the Botanical Gardens. On the last evening, the purpose of our visit having been completed, we were at a loose end and I suggested we take a walk along the boulevard that runs past the hotel and the Botanical Gardens. We had not gone far when our attention was attracted to some rats on the bank of a lake within the Gardens. Leaning over the stone balustrade we could look down about twenty feet to the bank below, and see the rats feeding and running about. We also noticed that every now and then one or other of the rats would swim across the lake to a clump of bamboo on the opposite bank, the journey by water being about fifty yards. The swimming rat would disappear in the tangle of bamboo leaves that hung over and dipped into the water; then we would see leaves higher up the bamboo moving as the rat climbed the stems and presumably played in there. This, however, is pure speculation, although it is the only interpretation I can put on it. From time to time a rat would swim back across the lake to the near bank.

To one side of the lake was a tiny island just large enough for mallard to nest on. Whether the rats had interfered with the ducks, their eggs or their ducklings, and there was some kind of feud between bird and mammal is open to question. At all events, as the first rat swam back from the bamboo clump a mallard drake entered the water from the island, swam purposively across in a direct line and stationed himself precisely at right angles to the route the rat was taking. There he remained, motionless, his head turned in the direction of the rat, apparently watching it. At that moment I said to Jane: 'Let's watch this closely. If the rat does not alter course it will bump into the drake and then we may see something interesting. The drake may attack the rat or vice versa. In any case there may be a fight.'

The rat kept going steadily until it was within a few inches of the drake. Another inch or two and it would have been within reach of the drake's beak. Then, suddenly, the rat dived without a splash, swam under the drake, reappeared the other side of it and continued swimming until it could land on the near bank. All this time the drake was still looking in the direction of where it had last seen the rat before it dived. It looked very much as though the drake had not got the wit to work out what had happened; and it seemed to me a very nice example of the difference between the nimble wits of a mammal and the relatively dull wit of a bird.

Perhaps we are apt to underrate the potentialities of ducks. They may look stupid at times. They may do stupid things at times. They may have absurdly small heads, and therefore absurdly small brains, by comparison with the bulk of their bodies. Their way of life may be largely influenced by Pavlovian reflexes. Yet they are by no means entirely stupid, and they are certainly not lacking in courage.

As to their mental capacity, there have been several recorded instances of ducks helping a disabled companion, which is not so surprising since they are nearly related to swans and to geese, the latter being more especially credited with intelligent actions. I recall a letter sent to me some years ago about an episode involving the whistling teal. The writer told how he was 'fishing on the bank of a broad river in Assam and flushed a pair of birds, one of which, obviously crippled, dropped into the water about midstream. Its companion dived under it and lifted it to a height of about ten feet, so that it was able to spread its wings and reach the opposite bank.'

The parent duck, in my family of thirteen, had a first brood, of which none survived. One by one the ducklings went. Some were trodden on and crushed to death by their female parent, and she accidentally drowned two by sitting on them in the water. But the rest were killed by rooks and rats. By contrast, her next brood survived intact, but there may have remained with her a memory of her previous misfortune. That is, however, pure speculation. All we know for certain is that one morning, when the family of thirteen had been penned up for the night and were let out, they had to step over the corpse of a brown rat of ample proportions. By careful examination of the scene of the calamity it seems very clear that the rat had ventured into the pen and had been trodden to death. It had entered through a small hole in the wire-netting covering the door of the pen, and its carcass lay flattened and mangled, but without lacerations of the skin, just inside this opening, its head pointing inwards into the pen. In view of the strength of a rat, of its savage disposition and its ability to defend itself with its teeth, it must have been met with an onslaught that gave it no time to retreat or defend itself. This may have been led by the aggrieved duck, with memories of the loss of her previous brood, or by a combined attack from the whole family. It could hardly have been because the mallard had stampeded or the rat would have got farther in than it did. For me this result was surprising.

Perhaps there was no need for this surprise, for in *The Countryman* for 1955 there is a note from Mr F. E. Welchman of Wiltshire about the rats in his hen run which his dachshund started to catch. Then Mr Welchman noticed there were rat corpses with wedge-shaped wounds on their bodies. 'The mystery was solved when, the next morning, the dachshund flushed a rat from a feed hopper and seized it. Before the dog could kill it nine hens started a "fearful shindy", rushed at the dog and snatched the rat from his mouth. The hens tossed the squealing rat from beak to beak and finally finished it off with savage stabs.'

Each year, in the early summer, our mallard drake began to look sorry for himself. The colourful plumage he was wearing a few weeks before would be gone and he would be beginning to look more like his mate, the duck, with her mottled brown plumage, although there would still be shades of his former glory. There would be, for example, vague traces of the green on the head and neck, but these only added to the general air of being unkempt, although neither that word nor any other in general use adequately describes his appearance. Much of the plumage was still neatly arranged, although patches of feathers here and there were ragged. Sometimes a bird will, in the course of the moult, show a more marked appearance of raggedness in the plumage, and in all there is at such times an air of depression, of lassitude and loss of spirits.

The moult in ducks is unlike that of most birds because the males go into what is called eclipse plumage. That is, they lose their distinctive colours and assume a plumage more like that of the females. The mallard drake will continue in eclipse until the end of August or the beginning of September, after which he moults again and his colourful plumage will be restored. The actual times vary slightly with the individual as well as with the area in which it is living.

There is another unusual feature in the moult of ducks. In most birds the loss of feathers is gradual and follows an orderly sequence. Not only is a balance maintained between the loss of the feathers and their replacement but the pattern of change is the same on both sides of the body. This not only keeps the body insulated against extremes of temperature but there is no marked loss of ability to fly and no disturbance of the balance of the body during flight. In ducks and their relatives, the swans and geese, however, as well as in flamingos, the large flight feathers are lost more or less simultaneously and the birds are for a while flightless. Ornithologists see in this a correlation between the eclipse plumage and the need for protection during a period of helplessness. The duck, it is argued, in her more sombre plumage, is less easily detected by predators than the more colourful drake. Certainly, a duck sitting on her nest amongst herbage will readily escape the human eye. The drake, now flightless because of the moult, and therefore less able to escape by flying, has the advantage of the mottled browns of the eclipse plumage and is less readily detected by a bird of prey.

Presumably, any protection in the camouflage of the duck's plumage and the eclipse plumage of the drake must be from flying predators. This much is suggested by a mystifying event that occurred a few years after we had become established at Weston House. The event was so baffling that I kept no written notes of it, but the main features are fairly firm in my memory.

One of our mallard ducks was sitting on a clutch of about a dozen eggs. She was there as usual one evening as night was falling. The following day, early in the morning, the nest was empty. The eggs, which were on the point of hatching, were gone – and so was the duck. Our gardener, Mr Fry, first discovered the loss: he saw the empty nest, noticed a few feathers leading to a spot about twenty feet from the nest and then discovered the body of the duck, more or less hidden among some scrub. He did not disturb the duck, which seemed to be so obviously dead, but reported the incident to me about an hour later and led me to the spot. The duck had vanished.

We stood looking at the spot where the body of the duck had been. It was marked by a feather, and a trail of small feathers led to it. We were discussing what animal could have killed the duck and carried off the eggs; and at that time I had no knowledge of the fact that a fox will remove eggs and bury them one by one, well separated. Since that time there have been several occasions when a visiting fox has buried eggs in the garden, just under the surface of the soil, in some instances incompletely buried so that part of the egg was visible. On one of these occasions the eggs could be identified as bantam eggs – they had a particular colour pattern – so that we knew they came from a neighbour's garden, where a bantam laying these peculiar eggs lived. This meant that the fox had travelled at least a hundred yards with each egg and scaled two walls, each not less than five feet in height, in doing so.

Then came the startling sequel to the disappearance of the 'dead' duck. While we were discussing the mystery of the vanished bird we heard a quack. It came from our duck. She was walking about near her nest, apparently none the worse for her adventure. We had failed to see her until then because the nest was in the middle of a slope, covered with abundant vegetation, with a shrubbery adjacent to it.

Shortly after this, there appeared in *British Birds*, for 1964, an account of mallard on the frozen surface of a flooded gravel pit being chased by two dogs. The ducks were evidently weak from starvation because, instead of flying away, they tried to escape in a zigzag mixture of weak flight and running. Then a drake became separated from the rest; he was buffeted and rolled over by the dogs. Finally, the drake lay motionless on its side with its head tilted back, whereupon the dogs lost interest. Once the dogs had departed the bird came to life again, struggled to its feet, shook itself and joined the rest of the flock.

At a later date somebody reported finding a duck buried alive, with only its head and neck visible above the ground. A fox was suspected and the mystery was solved when two scientists in the United States investigated the methods by which a fox attacks a duck. Incidental to their studies they found that if a fox steals up on a duck from behind and touches it with its paw or its snout, the duck makes no attempt to escape but goes into a sort of hypnosis or death-feigning, lying on its side with its wings tight to its body, its head and neck stretched forward and the legs stretched out behind. It may hold this position for a quarter of an hour or more. Should the fox wander away the duck may raise its head to look around but if it sees the fox it will slowly lower its head to the ground again. In this situation the fox seldom bites the duck, or does so only lightly, and it may even cache the 'carcass', digging a hole with its front paws, pushing the carcass into it, then shovelling earth onto it with its snout, in the typical cacheing action of a fox.

Something of this sort must have happened to our mallard and it is my regret that I did not keep detailed notes at the time. Fortunately the episode was not repeated and the mallard continued to be with us until the first time Jane went abroad and it was necessary to scale down the number of inmates of our zoo. There was no difficulty in finding a new home for the mallard. They were released on the Tillingbourne to join the ducks already there that had survived the unhappy incident referred to early in this chapter.

12 'That damned parrot'

Bassie was a parrot, an African grey. She first came to us when she was a year old and already an accomplished 'talker'. Her owner was due to go abroad and, reluctantly, had had to find a new home for her. From the moment she entered the house it was like another person added to the household, such is the gift of speech.

For me there was an added bonus. As I mentioned in chapter nine, I had become interested in studying the finer details of how a bird learns to talk and whether it merely repeats words like a tape recorder or whether, as when a child learns to speak, it associates words with objects or events. It soon became clear to me, beyond a doubt, that some parrots at least do more than repeat words mechanically, that when we speak of 'repeating words parrot-fashion' we do these birds less than justice.

First I analysed which words a 'talking bird' will repeat. For example, I soon found that Bassie would repeat a word after having heard it only once, although this was rare. More commonly, she needed to hear a word or a phrase repeated several times before she herself used it. Then there have been words or phrases which one person or another in our household tried hard to teach her, repeating each over and over again to her, yet never once did she attempt to say them. Such words are, so far as one can tell, no more difficult than the many others she did say. It is true parrots have difficulty with some consonants and vowel sounds, but even this does not explain their frequent failure to say words which someone deliberately tries to teach them, words which do not differ substantially from those the birds will repeat without prompting.

A phrase Bassie had learned before she came to us was 'Have a grape'. She said this whenever she was offered a grape, and would also alter the inflection as if asking for a grape. She would say these words the moment anyone took a grape from the sideboard, and later would say it if anyone merely walked to where the grapes were. In due course she was given apple or banana and she used this same phrase for these other fruits, so the phrase came to indicate 'fruit'. Moreover, she would say 'Have a grape' when the seed was put into her pot inside the cage, so the phrase also came to indicate 'food'.

We all talked to Bassie frequently and conversed together within her hearing, so she became familiar with our individual voices and would copy them, but not in a uniform manner. For example, we all said 'Hello' to her many times daily, as a greeting; and when one or other of us entered the room where she was she would say 'Hello', unmistakably in the voice of that person; but the performance was uneven. Thus, she mimicked my wife's voice seldom, and even more rarely the voices of Richard and Robert. She copied Jane's voice more frequently but less than she did mine, although Jane had more to do with her.

The association of voice with person was most marked. I had only to show myself in the open door of her room for Bassie to say 'Hello' in my voice, or to laugh exactly as I do. However, more significant were those occasions when she failed to get it right. Sometimes she would say 'Hello' repeatedly, in a variety of voices, as if she were not sure of herself and was having to take a number of samples out of the recesses of her memory to match voice to person.

A further example of association concerned the use of the telephone, and again the use of the word 'Hello'. For several years, in the house in which we now live, Bassie was kept permanently in the hall, near the telephone. In a very short while Bassie would say 'Hello' as soon as the telephone bell rang. The telephone bell is on the wall, near the ceiling, five yards away from the instrument itself. It was evident that she soon associated the two. She would say the word when someone lifted the receiver to initiate a call, in which case the bell did not ring, and she would also say it if someone merely walked in the direction of the instrument, without necessarily going to use it, and while that person was still several feet from it. So in this the parrot was recognizing several different situations; the bell ringing, the act of lifting the receiver, and the act of walking towards it. The fact that all were denoted by the same word 'Hello' is neither here nor there. In the same way I have known a small child call a telephone 'a hello'.

When one of us lifted the receiver Bassie would say 'Hello' in that person's voice. If nobody was in sight when the telephone bell rang she would use my voice, probably because I was then the most frequent user of the telephone. The most spectacular instance of association occurred, however, when I first had a resident gardener.

Wally and Phyllis Fry took up residence in the top storey of the house in the early summer of 1959, and during the rest of that year Wally spent nearly all his time in the kitchen garden. At midday Phyllis would go to a small window at the top of the house, thirty feet above ground level, and call her husband in to lunch. At such times her hailing cry, 'Wally', would be heard although she herself would not be visible from below.

It might be anything up to three minutes after his wife's clarion call had resounded across the garden that Wally would be seen coming down the path towards the house. During the first weeks of their residence Bassie had been in Jane's studio, which was directly below the window from which Phyllis called. Even with the doors and windows of the studio open Bassie would have heard the call, but indistinctly, and it would have had a non-directional quality, as the sound of an aeroplane overhead seems to come from nowhere in particular when the plane is hidden by cloud. Nor, from her particular position in the room, could Bassie have had a clear view up the garden path. (I tested both these things critically when the need arose to be sure of them.)

Certainly, Bassie gave no indication of having heard or seen anything until, during a spell of fine weather, her cage was put outside, directly beneath the window from which Phyllis called 'Wally' and with the length of the path in full view. After a few days of being out in the garden Bassie began to call 'Wally' whenever he came into her view, whether he was in the garden or in the house, and even when nobody had recently called him by name. And she seldom made a mistake. Unless there was a remarkable coincidence here, or I had overlooked something, we have to assume that the parrot, having heard a name called, had associated with it the physical appearance of the person who responded to the call a short while later. So she was associating a name with a person in the same way as we do.

After the warm spell had ended Bassie was taken indoors again and her cage was then put in the hall, at the other end of which was the front door. A day or so later Wally came in through the front door, which is something he does rarely. On seeing him the parrot immediately called 'Wally', in his wife's voice, of course, although she was seeing the man in an entirely different setting. For one thing, the hall is dark and anyone entering the door is silhouetted against the light from outside, so Wally was completely divorced from the circumstances under which she had first learned his name.

One thing that made these deductions the more reliable was that none of my family nor I had ever called our gardener anything but 'Mr Fry', so we could not have helped the parrot, even inadvertently.

Although this was the most spectacular example, it was by no means a solitary instance. Another directly concerned myself, so that I was able to keep a more careful tally of events. It began when Bassie was first placed in the hall, just beside the door into Jane's studio. When Jane was in her studio she might wish to say something to me without stopping what she was doing. So she would call 'Pa'. My study was forty feet away from the studio, and to talk to Jane I would get up from my desk, go through the door into the hall and say 'What is it?'

In a very short time, whenever I stepped into the hall from my study for any purpose whatever, whether Jane had called or not, Bassie would call 'Pa' in a near-perfect imitation of Jane's voice. As with the Wally incident, we have a name being called, a person putting in an appearance at a distance after a brief lapse of time, and the parrot linking cause and effect. A human being could do no better. In fact, he would do worse because he would not imitate so perfectly the sound of the caller's voice!

To a lesser extent, perhaps, learning the names of our dogs falls into this same category. You call a dog by name, either to give it a command or to summon it to you. In either case, the object is in one place, the source of the sound is some way from it. In those days we had our boxer-cross, Jason, and two Sheltie bitches, Suki and Poppet. Bassie would call Jason or Poppet by name, but never Suki. This may have had something to do with difficulty of pronunciation, or it may have been that Suki, who was Poppet's mother, was such a staid old matron we rarely had to admonish or command her.

Jason and Poppet were rumbustious and somebody or other was forever having to call them by name, reprovingly, as they raced about together. Moreover, both barked frequently and had to be commanded to silence. If they were racing across the lawns barking Jane would whistle them.

Bassie learned all this, except that she never attempted to imitate Jason's deep-throated bark; it would probably have been beyond even her capacity. But she often deceived us by her imitation of Poppet's high-pitched yap. She also called them by name, and never called Jason when only Poppet was present, nor vice versa.

It was not uncommon to see Jason or Poppet racing across the lawn and to hear Bassie whistle like Jane. The dog would half-prick its ears, give the slightest indication of checking its speed; and then immediately the ears would drop again as the dog continued to run at full speed. To us, used to seeing it, it seemed as if the dog, momentarily deceived, was saying to itself: 'Oh, it's only that damned parrot!' It is an indication of the superior power of discrimination of a dog's ears. I could not tell whether it was Jane whistling or the parrot, and was often deceived. The dogs were only momentarily deceived, perhaps not even that. Probably the quick lift of the ears was no more than a reflex, a response to a seemingly familiar sound.

All of us in the house were repeatedly deceived by hearing somebody calling, only to go along to find it was the parrot. This must be a common experience among parrot-owners, but not everyone lives in a big, rambling house with almost a Sabbath-day's journey from one room to another. We all must have walked many unnecessary miles during the years Bassie was with us because we could not tell whether it was someone calling or just 'that damned parrot'.

After a while Bassie invented a game, which she continued to play so long as the dogs were alive. She would bark like Poppet, then whistle like Jane calling Poppet in, and follow this by saying 'Poppet' in scolding tones in Jane's voice, without either dog or Jane being anywhere near. It was a perfect imitation of Jane calling a disobedient Poppet in from the garden. And to cap it she sometimes concluded with a rollicking laugh! It may or may not have been pure coincidence that Bassie, after re-enacting a scene, such as Jane calling Poppet in from the garden, should have ended with a rollicking laugh, as if herself enjoying the joke.

Mrs M. Reid-Anderson told me once about a young red macaw she had as a pet in what was then British Guiana. He spent part of his day in a custard apple tree, where a tin of maize would be hung for him to eat. Having consumed as much of the maize as he needed he would call the chickens and drop the rest of the maize to the ground for them to eat. 'Later, he would call them, and when he had got them all under the tree, with nothing to give them, he would scream with laughter!'

In addition to using my eyes and ears to make observations on Bassie's behaviour I was always on the look-out for other ways of testing what she did. The opportunity eventually came by accident, after we had had Bassie for about eleven years. Right from the beginning she had, as I am sure many other tame parrots have had, the habit of making a bubbling sound in imitation of water being poured. From the start she did this whenever anyone filled her water pot. At first we used a small jug and very soon Bassie would start to bubble as soon as she saw anyone approaching her cage carrying this particular jug. Then she started to bubble whenever she saw somebody carrying a jug that looked moderately like the original one. Over the years there were considerable changes in the water carriers we used. For a time a glass jam-jar was used. This was later replaced by a small watering-can of the kind used to water pot-plants. Then an old metal teapot was used. As a result Bassie was likely to 'bubble' at any water container being carried towards her cage or even one merely being carried about the room.

There came an evening, however, when I went to a dresser in the corner of our sitting-room, took from it a bottle of lime juice and walked across to the sideboard to get a tumbler. As I did so I heard Bassie, in the hall outside, bubble. This struck me as remarkable and I wanted to be very sure this was not due to coincidence. So I took careful note of the situation. The door itself was open about a foot. A line drawn from Bassie's cage through that opening intercepted my path from the dresser to the sideboard, towards which I was walking with my left side towards Bassie and the bottle in my right hand partially screened by my body. Re-enacting this showed me that Bassie could have had only a fleeting glimpse of the bottle, possibly of only part of it, yet she had recognized it as a fluid-container. At some time she may have seen fluid being poured from a bottle, but I can be reasonably certain a bottle was never used to replenish her water-pot. So I made a number of tests.

First, I allowed ten minutes to elapse, keeping out of the parrot's sight, then I re-enacted the scene with the bottle of lime juice and again heard Bassie bubble. Another rest and another re-enactment and again there came the bubbling. I then poured liquid from the bottle into a tumbler; again the bubbling. I substituted a wine glass for the tumbler; then I put away the bottle. I tried drinking from the glass. Each of these, repeated several times, with a short period between, evoked the sound of water being poured – the bubbling, as I have called it.

An objection to this could be that this was coincidence, because all talking birds — and Bassie is no exception — tend, like children, to have spasms of repeating one word or one phrase. It was noticeable that, during a single day, among Bassie's other miscellaneous jabberings, when she would go through her whole repertoire, one phrase would recur more than any other. It might go on like this to such an extent that, if you were compelled to be in the same room as the parrot, you would get sick and tired of hearing this one sound. Or she might repeat the phrase for an hour or so more or less continuously. Then, for weeks or months, she might not utter it again.

I had to be sure that on this evening Bassie really was responding to the sight of a bottle, tumbler or wine glass and not just having a 'bubbling fit', so that she would be likely to bubble whatever I did. That was why I allowed a reasonable period of time to elapse between each test. If, on doing this, she never failed to bubble at the appropriate moment, and also did not bubble at the wrong moment, then I could be justifiably convinced that she was reacting suitably to my tests.

During that evening and at intervals over succeeding days, I tried all possible combinations. I drank from a tumbler, a wine-glass or a bottle within sight of Bassie. I would lift an empty tumbler or wine-glass to my lips or merely carry either of them around, or carry a bottle, glass jug or china jug past her cage. Any action associated with the pouring or drinking of a liquid, or even merely putting my hand to a bottle, jug or jar elicited the bubbling. The tests satisfied me that this onomatopoeic sound used by the parrot, which for brevity I have called 'bubbling', signified a recognition of several different objects and actions: containers or carriers of fluid and the act of pouring fluids, drinking vessels as well as the act of drinking.

If we compare this with human speech two points can be made. First, much early human language must have been onomatopoeic, and doubtless in those early days one word may have served for several things, as in Bassie's bubbling. Secondly, children learning to talk — and even adults — often use one word to mean many associated things. For example, a man goes to a pub to drink (the act of imbibing fluid); on arrival he orders his drink (fluid in a glass); on returning home he may carry drinks (fluid in bottles) with him. In World War II an RAF pilot bailing out over the sea was said to have come down into the drink. Listening to an orator we drink in his words. And there must be many more such meanings for this one word.

While I was pondering these many tests, wondering about their implications, there came unexpected support for my tentative conclusions. Bassie was still quartered in the hall and Jane was in Africa. Until she departed, Jane had been largely responsible for looking after the parrot, giving her seed and fruit each day and keeping her drinking pot filled. I had undertaken to do these things during her absence.

Another routine procedure for whomever was looking after the bird was the goodnight ceremony. In this, you went towards Bassie saying 'Goodnight, goodnight' and this was the signal for her to go into her cage. You then shut the door of the cage and threw a square of cloth over the top. This last was traditional and probably unnecessary, but for Bassie it had become an essential part of the ceremony. The following morning the procedure was reversed, to the accompaniment of 'Good morning, Bassie', repeated throughout the ritual of removing the cloth, opening the door of the cage, filling the seed tray and replenishing the water.

One morning, working in my study with the door slightly ajar, I became dimly aware that Bassie was making the bubbling sound with unusual frequency. Being immersed in my work I took little notice of this at first, except to think it was a little odd. In time, however, this constantly repeated sound of bubbling water penetrated my brain sufficiently to make me stop work and wonder why. And then I went out to investigate. Bassie was perched on her empty drinking-pot. I had uncovered the cage, opened the door and filled the seed-pot but I had forgotten to fill the water-pot.

Having replenished it I waited and watched. Bassie drank and drank and drank, until the pot was two-thirds empty. The full drinking-pot normally lasted her a day. For her to have drunk two-thirds of her daily ration at one go, Bassie must have been a very thirsty parrot who had used the only means available to her of calling my attention to her needs. As near as makes no difference the parrot had used this vocal symbol, the bubbling sound, to express the concept 'water' – and almost, by implication, the abstract ideas 'empty' and 'thirsty'.

It must be confessed that, despite this salutary lesson, on three more mornings during the months that followed I failed to notice that the drinking-pot was empty. Each time Bassie drew my attention to her plight by repeating the bubbling sound at intervals of a few seconds until I attended to her needs. And each time, after the pot had been replenished, she drank about two-thirds of the water without intermission.

Some time after the events I have described I was refilling Bassie's seed-pot. Usually I plunged my hand into the large glass jar containing the store of seed and just threw a handful into the seed-pot. That morning, for no particular reason, I held the seed in my cupped hand and allowed it to flow slowly into the seed-pot. Bassie bubbled. I repeated this deliberately a number of times during the succeeding weeks. More often than not, when I let the seed flow from my hand, Bassie would bubble as she would at sight of a fluid being poured.

However, the best test, in my opinion, came when I filled a metal tube pipette-fashion and held one end towards Bassie's beak. She seized this as she would seize anything held towards her. As her beak gripped the tube I let a little water trickle onto her tongue. She had not seen the tube before. She could not see the water inside it. All that happened was that a little water touched her tongue, yet as soon as it did so Bassie made the bubbling sound.

One morning I was again immersed in desk work in my study when, almost like coming to from an anaesthetic, I became aware that Bassie was saying 'Goodnight, goodnight' at approximately two-second intervals. At first I merely thought it odd, as I had done when she bubbled interminably. Then, suddenly recalling my previous experience, I went out to her. I had forgotten her entirely this time. The door of the cage was still shut, the cloth was still covering the upper half of the cage. Bassie was hanging head-down from her perch, peeping out from under the edge of the cloth, and saying repeatedly 'Goodnight, goodnight'. It is perhaps not over-stating the case if we say that the parrot was using the only means available to her to demonstrate that, for her, and owing to my own forgetfulness, night time conditions were still prevailing at 11 o'clock in the morning.

A corollary to this came several years later. Jane, after visits to Africa, Malaya and Barbados, returned home and settled down to work again in her studio. Bassie was once again an inmate there, and soon her cage became the geometric centre of a ring of aquaria that decorated the walls of the studio, as Jane developed a new enthusiasm for keeping tropical and other freshwater fish. Closing down the studio at night was no longer a simple matter of saying goodnight to Bassie, closing the door of her cage and draping a cloth over the cage. Each aquarium was illuminated by its separate light and these had to be switched off one after the other.

One night Jane was struck by the way Bassie, who had been put to bed first, continued saying 'Goodnight, goodnight', as she made her way from aquarium to aquarium, switching off the lights. She then found she had omitted to shut the door of the cage. Subsequently, a similar thing happened. This time Jane had forgotten to put the cloth covering over the cage.

It was possible to stimulate Bassie to continue calling 'Goodnight, goodnight', merely by leaving either the door of the cage open or by not putting the cloth over the cage. It was her method of drawing attention to an uncompleted ritual. So while it may be true, as some people claim, that parrots cannot think and have no idea of the meaning of the words they utter, they do something very near to both of these. They are aware of something missing, or of an unusual situation, and they seek a remedy by using the most appropriate mimicked sound in their repertoire. Moreover, once the appropriate action is taken, the bubbling sound or the word 'Goodnight' ceases. And this, it seems to me, amounts to much the same thing as elementary thought. At the very least, these tests suggest that talking birds sometimes show association of words with objects, persons or situations.

There are grounds for supposing that talking birds sometimes show signs of practising their mimickings. One writer, at least, has suggested that when a parrot suddenly says a word perfectly for the first time two weeks after it has heard it, it has been practising the word in its head. This is not unreasonable, although hard to prove. There is more evidence from words or phrases practised out loud.

Soon after Bassie first came to us Jane tried to teach her to say her own name and to repeat our telephone number. The idea was that if the parrot happened to take wing one day – we preferred not to clip her feathers – anyone finding her would soon be able to learn where she belonged.

In a short while the parrot mastered the phrase 'I'm Miss Bassie bird'. Then came the next step, of trying to teach her our telephone number which was then East Horsley 2452. She soon learned to say 'East Hor'. Then she managed to say 'East Horsley'. The numbers she found more difficult. By dint of much patience on Jane's part Bassie managed to say 'two four'. The 'five two' seemed to be beyond her. In the later stages of her tuition she would confidently say 'East Horsley two-four' and then, as if in desperation, would repeat the 'two-four' interminably. Either that, or she would repeat the phrase 'East Horsley two-four' over and over again, finally breaking off and pouring out a torrent of other sounds from her vocabulary, in a jumble of words, mechanical sounds, whistles and what have you, as if in sheer exasperation.

Bassie certainly practised on her own. Had it not been amusing to us she would have driven us crazy with the constant repetition of 'East Hor . . . East Horsley . . . East Horsley two-four, two-four, two-four, two-four'. It reminded us of that tantalizing situation in which a beginner on the piano goes over a phrase of notes interminably, always getting stuck on the same difficult chord.

For a while, then, we heard 'I'm Miss Bassie bird, East Hor . . .' or 'I'm Miss Bassie bird, East Horsley two-four, two-four, two-four, two-four', and all manner of combinations of these words. Then the rot set in. She finally settled for 'I'm Miss bird', sometimes varying with 'I'm Miss Hor bird'. And at times she would follow this with a long string of 'two-four, two-four, two-four, two-four', delivered in such a rollicking tone that it sounded like a peal of derisive laughter.

It so happened that, during this phase of Bassie's education, a staid elderly lady, a Miss Bird, was coming to visit us one afternoon. We were in fear and trembling about what might happen. Fortunately, Bassie did little more than sit on her cage and look glum throughout the visit.

By contrast with the unsuccessful efforts made to master what was admittedly a difficult exercise, there was another occasion on which the same practising led to perfection. It is less noteworthy but is more in line with what must be common experience on the part of parrot-owners. It also illustrates the converse of being taught, the parrot's ability to teach itself.

It took place around about the same time as the 'East Horsley 2452' episode. My brother-in-law was staying with us at the time. He was passing Bassie's cage one morning when she, perched as usual on top of her cage, indulged in her familiar trick of lunging with her beak and hooking it in his clothing. This is something of which she never tired, and it kept us always on our guard for fear that either she might lunge at our faces or that, with her beak hooked in our clothes, we might pull her off her perch by not stopping quickly enough.

My brother-in-law, finding himself held by a beak hooked into the wool of his cardigan, said jokingly: 'I'll wring your neck'. The phrase seemed to tickle his fancy and for the next few days, whenever he passed the cage, which was about half a dozen times a day, he repeated (parrot fashion!) the remark 'I'll wring your neck'.

Bassie merely gave him the usual deadpan look, of which parrots are past-masters, and made no response. Then we heard her saying: 'I'll wring . . . I'll wring . . .' as if the complete phrase was beyond her. Then suddenly, out it came. 'I'll wring your neck.' From that moment onwards she never looked back. The phrase became part of her repertoire, to be used occasionally when she was in full voice and running through everything she had learned. But more especially, and even after the passage of years, she would use it when she had made a lunge at somebody's jacket or finger. And always, of course, it was said in my brother-in-law's voice.

There was a feature common to these two incidents: that in neither of them was there the simple reproduction of sounds as in a tape-recorder, perfect at the first playing and never varying once the recording had been made. There was repetition in order to reach perfection. There was the faltering, the temporary failure and, in the case of the telephone number, a final failure to reach full achievement. At the very lowest, therefore, we can assert that as between the tape-recorder and a parrot's mimicry the first is automatic and constant and the second is unpredictable and has plasticity. It is the kind of plasticity a child exhibits when first learning to speak, or even that an adult often shows when trying to master a difficult word for the first time – or when learning a new language.

With regard to 'I'm Miss Bassie bird: East Horsley 2452' there was another important feature. There was a mixing up of the separate parts of the two phrases, a re-sorting and transposition, much as a composer re-sorts and transposes a phrase of notes, varying his themes to produce a symphony. Moreover, the lilting two-four, two-four, two-four, two-four went far beyond Jane's attempts to teach the bird a simple sentence. We might, with a little exaggeration perhaps, say that the parrot was treating the words as a poet might, departing from straightforward speech to introduce a musical quality.

But Bassie went further – unless I am very much mistaken.

Our family is not given to whistling. None of us ever whistles except Jane, and she merely used a somewhat strident whistle to call back the dogs. So far as I can recall we have never had staying with us anyone addicted to casual whistling tunes. Bassie, on the other hand, has indulged in a variety of whistling notes. These may have been native to this kind of parrot or they may have been mimicked sounds picked up over the years. What was noticeable was that as the years passed she tended more and more to string these together. I was also struck by the fact that, when the house was quiet and I was sitting in the room next to the one Bassie was in, alone, with both doors open so that I could hear her plainly, she appeared to be making up tunes. By this I mean she was whistling notes consecutively that vaguely resembled a melody. It was the kind of whistling, nearly tuneless and almost aimless, that a child uses before its musical education is far advanced, or that some men indulge in when whistling absent-mindedly while engrossed in their work.

When at her best Bessie would prattle for half an hour or more on end, with a torrent of words, snatches of tunes (sung, not whistled), bird calls, barking dogs, mewing cats and a wide variety of mechanical sounds. Indeed, we can say here again, that there was no comparison with a tape-recorder, because these mimickings did not come out in orderly sequence. The torrent was varied from one repetition to another. The principle, it seems to me, is the same as in the whistling, a matter of the bird picking items out of the recesses of its memory and stringing them together: a kind of vocal dreaming.

There is always a danger of reading too much into such happenings, so one is constantly on the watch for supporting evidence that will prove acceptable to a scientist. In the episodes I've quoted there is the strong suggestion that this particular parrot was using not merely the so-called parrot-like repetition, to use the customary phrase, but was moulding the pattern of her learned sounds much as people use sounds in early speech, or even as a poet or, more especially, a composer uses sounds.

In early 1961, Jane was holding an old metal table fork used for mixing the animals' food, twanging the prongs with her fingers. After a while we became aware that Bassie was walking round the top of her cage rubbing the wires with her beak and so producing a sound very near to that coming from the fork.

When Jane is working in her studio, with cameras and floodlights, there are numerous metallic squeaks as nuts are turned to raise or lower the floodlights, or the rod holding a floodlight slides in or out of its tube. There are clicks from the camera itself, noises of boxes being moved or cages opened, and other trivial sounds connected with the work. I have, on such occasions, seen Bassie running around her cage tapping a wire here or a stouter piece of metal there, to copy the sounds she was hearing. The result is that one heard the sounds coming from the apparatus and, like echoes, Bassie's imitation of them. The only difference was that the 'echoes' were often louder than the original sounds and were repeated more frequently.

The main surprise was the speed at which these sounds, mimicked mechanically, were produced. It almost leads one to believe that Bassie had memorized which parts of the cage produced which notes, just as a harpist knows which strings to pluck. Indeed, I feel it is an inescapable conclusion: that the parrot was using the cage like a harp.

There was another occasion when Jane was cleaning and polishing a piece of plate glass. She needed to be sure its surface would be free of all dirt and blemishes, so she polished it vigorously. The soft leather produced high-pitched protesting squeaks from the glass and in a while we became aware that Bassie was also making only high-pitched sounds. We recognized each as one she had used on previous occasions.. Soon it became apparent that she was dragging each sound in turn from her memory and matching it against the high-pitched noises coming from the glass. In the end she was repeating a sound, one we had heard her use before but of unknown origin, which to our ears was a perfect match for the squeaks now coming from the glass. Having matched the sound with one she already had stored away she proceeded to use only that one, again acting as an 'echo' to the sounds Jane was producing.

If anything, this was the most incredible performance of all. Instead of remembering the sounds made by the various metal parts of the cage she was drawing upon her store of memorized sounds and using the voice to reproduce them. The difference between the two processes seems to me to be the same as between a harpist playing from memory and a singer singing from memory.

On another day when Jane was busy in her studio, Bassie was watching all her movements with an air of apparent interest. From time to time she mimicked the clicks and squeaks coming from the apparatus. Then Jane had occasion to pull out the sliding lid of a box. It made a harsh grating sound. Bassie did not copy this but produced a sound very near it, one with which she was familiar and which she had already perfected. It was the harsh, grating sound made by the tray in the bottom of her cage, as it was being withdrawn for cleaning.

This, on its own, might have been mere coincidence. The same can be said of some of the other episodes. It is when we put them all together that they acquire a greater significance. On several occasions we have had a joiner working in the house. Invariably, Bassie found a spot on the stouter metalwork of her cage that, perfectly or nearly so, resembled the sound of his hammer. Then she hit this with her beak in perfect time with the sound of the hammer, but infinitesimally behind each hammer-blow, imitating not only the sound but the rhythm.

Finally, Bassie had another trick which was, in a sense, the reverse of these actions. She would often strop her beak on the metalwork of her cage. Presumably, this was the analogue of a fruit-eating bird cleaning its bill on a branch. As Bassie stropped her beak on metal she mimicked vocally the sound it made, the real and the mimicked sounds synchronizing but with the latter much louder, so that the original sound appeared like an anticipatory echo.

In the summer of 1964, when Jane was in Barbados, I heard the sound of typing in Jane's studio. As I listened more carefully, I could recognize that not only was the sound that of Jane's typewriter but the manner of hitting the keys was hers too. She typed with two fingers, not very expertly, and that was what I could now hear: tap . . . tap, tap . . . tap, tap, tap . . . tap . . . tap . . . I confess I found this very eerie. Spook stories ran through my mind, of manifestations associated with a person far from home being followed by a telegram telling of a tragedy.

The studio door was wide open, so I walked on tiptoe to peer through the gap between the door and the jamb. The typewriter on the desk was still in its case, although the sounds of typing were continuing. Then I saw Bassie perched beside the water-pot in her cage, and there could be no question that she was the phantom typist. She was plucking with her beak at one particular spot on the metal band that holds the drinking pot, not butting with the beak, as is more usual. Having established this by watching her, I then tested the various parts of the cage and found that this was the only spot on it to give a correct representation of the sound of this particular typewriter. There seems to be no other conclusion, than that either the parrot had discovered by accident that this particular place on the metalwork simulated the sound of a typewriter, or she had found precisely the right spot by going round the cage searching for it. The fact that she had to perch in an awkward position to do so, and, moreover, in a position that she did not normally assume, makes the second more likely, incredible though it may seem.

During the days that followed, Bassie had a veritable orgy of typing: always sitting in the same awkward position and always striking with her beak at this one spot. As I continued to listen I could recognize how closely the parrot was imitating not only the sound of the keys on this particular typewriter but the particular rhythm or pattern used by Jane. There was another feature of it, but whether or not this was coincidence we shall never know. Jane, before her departure for the West Indies, had been typing perhaps throughout the day, or for half a day, or perhaps only for a few minutes during one day, as when typing a single letter. For several weeks Bassie followed the same routine. One day she would 'type' incessantly, from morning to evening. Perhaps the next day she would do so for a few minutes only, and then continuously for half a day on the following day. It was almost as if the bird, missing her constant companion, was trying to recreate her familiar presence.

It was during this period that we had a visitor. Not long after her arrival she used the telephone, which was just outside the studio. Seeing the door of the studio ajar, and having heard the sound of typing in that room, she put her head round the door, after finishing her telephone conversation, to say she hoped she had not disturbed whoever was typing. She saw nobody in the room, which momentarily made her flesh creep, but she soon discovered the sounds were coming from the parrot's cage. Her remark was that 'it sounded like somebody typing hesitatingly, as if thinking something out while striking the keys'.

I have no doubt that there have been and still are many owners of parrots that could match what has been said here of Bassie's ability to associate words and sounds with objects and events. I would go further and say that I have collected such examples of other people's experiences. Pride of place is given here to Bassie simply because, in writing about her, I am drawing upon my own first-hand observations.

Bassie's colourful companionship ended in tragedy after she had been with us fifteen years. Somebody had brought Jane a semi-tame stoat. She took it into her studio where it slipped from her hands and escaped up the chimney of the open fireplace. It was impossible to retrieve it. The studio door leading into the garden was left open and, after a lapse of hours, with no further sign of the stoat, it was assumed the animal had departed.

The following morning Bassie lay dead on the floor of her cage, with a typical stoat bite on the back of her neck. We never saw the stoat again.

I took the limp corpse in my hands, with a lump in my throat. Words failed me.

Mynahs and starlings 13

One morning in 1958 the telephone rang. It was Gavin Maxwell, famous for his book about otters, *Ring of Bright Water*, though it was then still to be written. This morning he was calling about a bird, a hill mynah he had recently acquired, and he invited me to call on him at his London home to see it.

Gavin knew that at that time I was collecting instances of animals playing with fire, to give it an omnibus term. His hill mynah was given the freedom of the living rooms. The moment anyone lit a cigarette the mynah would fly to it, bite off the burning end and swallow it. Naturally, I was all agog to see this and I hastened to his house. Gavin lit a cigarette, stood in the middle of the room, and almost immediately Kali, the mynah, flew past his face and there was no glowing end of cigarette to be seen, on the floor or in the bird's beak.

I wanted to be quite sure that we were not being deceived in any way, so I asked Gavin to re-light the cigarette while I stationed myself where I could watch events even more closely. Once the end of the cigarette was glowing red, Kali flew past Gavin's face. I was able to see the bird snap at the glowing end with its bill and then land on the back of a chair. The trick was repeated several times for my benefit and there was no doubt that the mynah was snapping off the hot end of the cigarette and swallowing it. What is more, it seemed to take no harm from this strange meal.

The following year Gavin announced he was going abroad and asked me if we would take care of Kali during his absence as well as another special pet of his, a Brazilian hangnest, a most charming little bird. Kali was placed in an aviary a short distance from the house and less than twenty feet from the high wall that separates the garden of Weston House from the main street that runs through Albury.

Kali was a most amusing pet and visitors to our menagerie, having made the round of our other aviaries and pens, would always return and spend a long time at his aviary. I recall particularly how my cousin Albert, a farmer more used to cows and pigs, who had never had a pet bird, would stand looking at the mynah for long stretches of time, an amused smile on his face. Kali seemed to fascinate him.

The hill mynah as a species has the reputation of being a great talker. Some people rate it higher than a parrot for mimicking human speech, but I would challenge that. At all events, Kali's repertoire was strictly limited. His favourite saying was 'Hullo, Kali', and he usually followed this with a loud penetrating wolf whistle. Other than this, he frequently laughed (Gavin's laugh) and coughed (Gavin's cough) and at times he would mutter in low tones in what might have been Hindustani, Urdu or merely garbled English. We never could make any sense of it. Any visitor who had lived in India was asked to listen in the hope of getting a clue, but it was all to no avail.

Then, one day, we found somebody had left the door of Kali's aviary ajar. The mynah was gone. We searched the garden, and the fields and woods beyond. Gavin had entrusted his prized pet to us and we had betrayed that trust. What, we asked ourselves, would we say to him when he returned home and asked for his pet?

By chance, a few days after Kali's departure, we saw an advertisement in a newspaper offering a young hill mynah for sale. It entailed a long journey and a quite high price, but we were desperate. We hoped, in a futile kind of way, that Gavin would perhaps not notice the substitution. Drowning men, we are told, clutch at straws. The journey was made and within an hour of the new mynah being installed in the aviary the telephone rang. The caller was a farmer's wife, living less than a mile from us, ringing to say that a week previously a strange bird had come into the house. She had put it in a cage and fed it, and then someone had told her that we kept birds and it might belong to us. Oh joy! The strange bird proved to be the prodigal Kali!

There were now two mynahs in the aviary. One, the young bird bought recently, had uttered no sound during the hours we had had it. The other was the loquacious, laughing, coughing, muttering Kali. Within a few weeks, the newcomer had grown equally loquacious but its whole repertoire was precisely the same as Kali's: the same 'Hullo Kali', the same laugh, the same cough, the same wolf whistle and the same incomprehensible muttering. It was mimicking, but it was mimicking only its companion. In all the time we had them, neither bird added one new syllable or sound to its repertoire, even though we encouraged them to do so.

It was too absurd to see two birds in an aviary, looking as alike as two peas in a pod, and each talking, coughing, laughing and muttering in identical terms. When cousin Albert visited us, after the second mynah had been added to the aviary, he became almost rooted to the spot. Instead of merely an amused smile, we could look down the garden and watch him laughing aloud, all by himself. Neither mynah made any attempt to mimic cousin Albert's laughter.

The wolf whistle, as we all know, is that loud double-noted whistle, the first note ascending, the second note descending, in common use by quasi-amorous young men to attract the attention of a young woman. The inhabitants of Albury soon learned the meaning of the strange sounds that came from behind the wall bounding our garden. But many strangers walk through the village, especially at week-ends. I have often wondered how many of them, the young women especially, had been puzzled by the wolf-whistle when no one was in sight to give it – for it was most penetrating and could be heard distinctly two hundred yards away. There was one elderly woman who frequently passed beside our wall. She confided to me that, until someone had told her of our mynahs, she thought some cheeky boy was wolf-whistling her while hidden somewhere.

On Gavin's return he collected the Brazilian hangnest and told us we could keep the mynah. So we need not have been so bothered about finding a substitute when we thought Kali was irretrievably lost. It was, however, a lucky accident, for it gave us the opportunity for a most interesting study.

Mynahs belong to the starling family and starlings are wonderful mimics in both sound and movement. The classic example is of a starling that lived near a bell tower in which the bell was visible from the outside. This starling was seen to imitate the sound of the bell and to sway in time with the swaying of the bell.

I have no experience myself of starlings imitating the physical movements of other birds but I have it on reliable authority that a common starling was seen flying in perfect imitation of a swallow hawking insects. I have also been told of a member of the same species flying like a lapwing and of another flying like a green woodpecker. From my experience with the two mynahs I would not be prepared to reject such stories, and the evidence on which my credulity is based comes especially from the two Kalis, as we came to call them, since in their behaviour especially they looked more like twins than two totally unrelated birds.

It was noticeable that, in addition to the two birds having the same vocal repertoire, when one mynah preened the other would preen, when one went to the food bowl to feed the other would do likewise. Similarly, when one flew from one perch to another the second mynah would do the same. They seemed to copy each other in all their actions. I have seen something like this in the wild starlings that spent the day in the garden. It was even more noticeable in two grey parrots. It seems as if vocal mimicry tends to go hand in hand with mimicry in movements.

One of the two grey parrots in question was Bassie. The other was an African grey, known as Cockie, that we took care of while its owner was on holiday. They were housed in the same room, each in its separate cage and separated by about three feet. It soon became noticeable that when one drank the other drank, when one fed the other would go to its food container, when one preened the other preened. I would have liked to have studied this over a longer period but unfortunately the owner's holiday was not a long one.

The most striking example of this kind of mimicry was seen before Kali made his memorable escape, while he was still the sole occupant of the aviary. Someone brought me a fledgling jackdaw that seemed to have lost its parents, and I hand-reared it. Young jackdaws are relatively easy to bring up. They are little afraid of human beings and readily accept food, which may be the reason why they are often adopted as pets. This one behaved in the usual way, opening wide its beak as soon as it saw food and making the soliciting call characteristic of a fledgling jackdaw. At the same time it would half-spread its wings and flutter them in the usual supplicatory manner.

In books on birds it is customary to render the call of jackdaws as 'tchak' or 'chak'. It is practically impossible to represent the call in words, so for sake of brevity we will call it 'Jack'.

When I fed our fledgling it opened wide its beak, half-spread its wings and fluttered them, at the same time crouching so that its body came more into the horizontal. It also made the call which sounded something like 'Jack'. As it did so I pushed food down its throat with thumb and forefinger. As it swallowed the food its call became transformed into a gurgling, satisfied sound which bore a close resemblance to 'Jack' being repeated under conditions of partial strangulation.

This gurgling may well be essential, telling the parent that the food has been well and truly placed down the infant throat to continue its journey down the gullet to the stomach. Without some such guidance the parent bird might well risk having the packet of food placed only in the infant's mouth, to be immediately scattered right and left and so lost, as so often happens when one inexpertly feeds a young bird. When the human foster-parent allows food to be wasted in this way he can readily replace it. The jackdaw parent has to spend precious time laboriously searching for more in the fields.

In the natural course of events, the time came when the young jackdaw developed its swallowing reflex and was able to pick up food for itself. Even then, although it could quite reasonably take food from the bowl placed in front of it, the young jackdaw preferred to be handled. Consequently, it still continued to gape, crouch, flutter its wings and call 'Jack'. Because it pleased me to have a young animal showing such confidence in me, I continued, for at least part of its daily meals, to push food down the baby jackdaw's throat.

It was necessary also, at this time, to put the growing jackdaw into a vacant aviary, and the only one available was the one next to that inhabited by Kali. Soon I noticed that whenever I took food into the jackdaw's aviary, Kali would crouch, croak and move its body backwards and forwards in a jerky manner. On the first day I saw this I thought the mynah must be ill. It looked to be on the verge of epilepsy or some such affliction, and for several days I puzzled over it, for at other times Kali was obviously his normal healthy self.

I began to take more careful note and found that it was only when I went near the aviary with food that the mynah exhibited these distressing symptoms. Then it dawned on me. This seeming epilepsy was a caricature of the supplicatory actions of the young jackdaw. By careful listening one could detect in the croak a reasonable imitation of the jackdaw's supplicatory call. There was also a passable imitation of the jackdaw's satisfied gurgle. The jogging backwards and forwards seemed best translated as an attempt to mimic the wing-fluttering of the young jackdaw. Had the imitations, vocal and bodily, been more perfect I should probably have assumed the mynah was merely going back to its own infancy and reviving a pattern common to all baby mynahs. As it was, there seemed little doubt it was giving an incomplete or clumsy imitation of the young jackdaw's voice and begging movements. That stimulated me to search for similar examples.

I was feeding the jackdaw on bananas and grapes, both of which were the mynah's normal rations, and also on wholemeal bread mashed with honey, which Kali seemed to relish above all else. This may have given it the stimulus to do something positive in order to attract to itself more of the bread and honey. Some birds are born cadgers; gulls are a good example, sparrows another. Yet the fact that the mynah's 'epilepsy' ceased entirely a few weeks from the time I discontinued hand-feeding the jackdaw seems significant. It was never repeated after that time.

One advantage of having the kind of zoo in a garden we have at Weston House is that the wild species living unconfined and quite naturally within the grounds become, for purposes of study, part of the menagerie. At the time when the Kalis were with us – and they lived several years beyond the time of the jackdaw incident – there were plenty of starlings around. Several pairs nested in the roof-space of the house. Other pairs nested in hollows in the trees. And the topmost branches of a very tall lime tree on our boundary, more especially, served as an assembly point for their departure each evening to roost on the buildings in the nearby city of Guildford. So we had plenty of opportunity for studying them.

The starlings foraged on the lawns and bathed in the puddles that formed at the edges of the paths when it rained. We had ample opportunities for seeing this, for when one of them bathed all, or most of, the others would bathe. If one stopped while searching the grass for insects to preen or scratch the chances were that one or more of its fellows nearest to it would also stop and preen or scratch. And in the way all would fly up together into the topmost branches of the lime they showed far greater unanimity and synchronization of action than most species. This, indeed, is something many writers on birds have emphasized.

It was, however, the vocal mimicry of the wild starlings that was the most fruitful source of obervations. It was not unusual to hear a blackbird singing – only to find it was a starling giving a perfect imitation of the song. When a starling on the roof of the house gave a perfect imitation of a curlew we knew that at least one of our starlings had spent part of its life in a district where curlews lived.

We had a pond in front of the house with a fountain, which was worked by a hidden electric motor that could be switched on and off. It was small as fountains go, sending up a slender jet of water that rose about three feet into the air before losing momentum and returning to the surface of the pond in a hundred or so drop-

Bassie, the West African grey parrot, in the pose of a prima donna singing 'Oh for the wings of a dove'

A pair of young magpies
looking absurdly old and
mature: an outstanding
feature of them was the way
they copied each other's
movements

Kali, the hill mynah,
loquacious but with a limited
vocabulary and given to biting
the lighted ends of cigarettes
and swallowing them

The author with one of his
magpies

Sally, the aged and staid donkey, on the left, being confronted by her skittish two-year-old companion Minnie, short for Miniheehaw, an outrageous pun prompted by her braying

Sally, the elderly donkey, putting on the doleful expression that earned her many gifts of food from village children brought to inspect her

A baby moorhen in its fluffy black plumage showing the disproportionately large feet of a bird that prefers to run rather than fly

A five-day-old moorhen chick with its foster mother, a bantam hen. The chick expected food to be brought to it, the hen expected the chick to come when called. It was the moorhen chick that learned to break the deadlock

erce, the crocodile, as he began to grow up, was sometimes allowed his freedom to wander in the
arden but he spent most of his time prone on the ground only rarely moving about

aughter Jane with her son Mark and daughter Hazel enjoying the sight of their pet iguana

Weston House from the garden, showing the mock Tudor chimneys that are a feature of this part of Surrey

lets of varying sizes that produced a continuous tinkling, pleasant to hear.

One day I was standing on the steps leading to the front door when I became aware of this familiar tinkling sound. But the fountain was switched off. After listening for a few seconds, to make sure I was not merely 'hearing things', I looked up. As I suspected, there was a starling on the hip of the roof mimicking the sound. When one considers how many different tiny sounds go to make up what we call 'tinkling', this piece of mimicry represents quite a formidable vocal feat.

In the same category must go the performance of another of our wild starlings a few years earlier. For several weeks we had been hearing the drone and rasp of a power-driven saw. Woodmen were clearing a copse about half a mile from the house. First we would hear the sound of the saw, then the crash of the tree as it toppled over and hit the ground. These two sounds alternated throughout the day except for an hour at midday when the men went home for a meal.

One day, during the lunch break, I heard the sound of a saw and the crash of a tree, but not so loud as usual. When I went to investigate several small birds flew out of a beech just beyond my boundary. One, larger than the rest, disappeared into the branches beyond. At the same time I heard what sounded like a throaty human laugh. This pulled me up short wondering if I might have flushed some unusual migrant. It was, however, only a blackbird. The real culprit proved to be a starling that remained alone in the beech.

As I stood and watched it indulged in starling calls and alternated these with the familiar alarm notes and snatches of song of a blackbird. It managed also groups of un-starling-like sounds that at first were quite unintelligible, as well as the sound of a power-driven saw. In the end I realized what the unintelligible sounds were. They represented the numerous and different sounds of a tree falling to the ground. First there was the swish of the foliage followed by the snapping of twigs, the crack of the breaking branches and the thud as the trunk struck the ground, ending in the smaller sounds of the main foliage swaying before coming finally to rest. It seems unbelievable, now that I come to set down this episode on paper, and I might have some doubt about it except that I made notes at the time and these I now have before me.

It seems, however, that my experiences are not unique.

Although this has nothing to do with our private menagerie it seems worth while quoting from a letter sent me by Mr F. A. Hammond, of Lydd in Kent. The writer told how it was his custom to take a daily walk around the local football ground.

> One morning while taking my constitutional round the perimeter of the football ground I halted in amazement. The roaring of a cheering supporters' crowd seemed to come upon the air from an emptiness about me. There was no doubting it. The noise rose and fell as such noises do owing to the excitement of a scurry around the goal-mouth. No players, no ball, no crowd. Just me, trees, goal posts and snow plus wind with flying clouds overhead.
>
> After locating the direction from which the noise issued forth, I came upon one particular tree which had no bare branches and twigs but was loaded with black foliage – a foliage of starlings giving a chorus in bird mimicry of many human voices of different pitch and quality in a frenzy of excitement.

I regard this as one of the most striking examples of bird mimicry. One can only suppose that the starlings were indulging in the kind of mass chorus that can sometimes be heard, especially when birds are assembling to go to roost. This, however, bears no real resemblance to the noise of a supporters' crowd at a football match. Nor would it do so if only one or two of the starlings were imitating the appropriate human voices. The only conclusion seems to be that a fair number of the members of the flock were actually reproducing the sounds of an excited football crowd. If so, then we are dealing with the most extraordinary instance of mass mimicry yet recorded.

Mrs A. Shirley wrote to me in 1964, from Nova Scotia, about a baby European starling. This species was introduced into the United States in the early years of this century and has since spread over most of North America. Mrs Shirley's baby starling had been rescued as a fledgling, hand-reared and kept in a cage. It was named Stardust and when about nine months old it suddenly was heard to say, in a small voice, 'Go to bed, Stardust.' The starling used a number of other words and phrases. 'His voice was so tiny, yet so exquisitely distinct one might have imagined an elf speaking.'

This last prompts me to set down here a favourite theory of mine, that haunted glades and woods, even the beliefs about Little People (the elves and fairies), may sometimes have originated from superstitious folk hearing a wild bird mimicking human speech.

On the rare occasions when, being sleepless, I have got out of bed and walked downstairs, Bassie has never failed to say 'Hello', in a soft sepulchral voice. I have often wondered what the effect of hearing this would have been on a burglar, already nervously tense!

My pet theory about haunted places was born one day in 1954 when I was working quietly in the garden and 'heard voices', snatches of conversation, softly spoken, coming apparently from nowhere. I quickly realized that the words came from Jasper, the jay, hidden from view by a shrub.

Taking a further flight of fancy, I have sometimes wondered how many houses reputed to be haunted have gained their reputation because of talking birds. Jackdaws can be accomplished mimics, and they often build their nests deep down in chimneys, especially the large chimneys of big, old houses. Their flutterings can be heard through the walls at times, so why not their voices? Many birds use their voices at night, especially their subsongs. It only needs a jackdaw to utter a few soft words of mimicked speech, heard by one person in the dead of night, and human imagination would do the rest.

Another example of mimicking, similar to the non-vocal mimicking of starlings, came our way in the early summer of 1957. Our garden at Horsley then looked more like a typical zoo than ever. In addition to the permanent aviaries and pens, it seemed to be littered everywhere with cages, large and small, accommodating temporary visitors. This was the season when young birds were coming off the nests and tending to become casualties, either through wandering away from the rest of the family, being overlooked by the parents, or from some other cause not easy to determine. The occupants of the temporary cages also tended to draw free birds into the garden, solely because the cages contained food, which the growing youngsters were apt to scatter through the wires of their cages to the outside, so there were pickings to be had everywhere. A family of sparrows were regular visitors on the back lawn, where two of the cages stood. The two parents and five youngsters would come down from the roof and, while they were foraging, the lawn seemed to be alive with sparrows. The wonder was that the family kept together so well, and it may be that the antics of a couple of young magpies gave a clue to this.

These two magpies were brought in to us by a party of small boys – very small boys. None of them seemed to be very definite about where the birds had come from, but it appeared that each boy in turn had tried to persuade his mother to be allowed to keep them. So from the chorus of small voices it seemed to emerge that the magpies and their escort had experienced a long and disappointing peregrination, which ended on our doorstep. The birds were fledged, with little of their down remaining and their tails about half an inch long. It was unlikely, therefore, that they had arrived at the stage of voluntarily leaving the nest. They had no idea of opening the beak to take food, which is usual at that stage, so it was a matter of teaching them by forcing the beak open and pushing food down the throat. At the fourth lesson they began to get the idea, and at the sixth feed they gaped of their own accord at the sight of food held before them in forceps. Thereafter there was no further trouble; each threw its beak wide open the moment food appeared, or even if there was any kind of movement in the vicinity of their cage.

In learning to feed, young birds are helped at first by simple inborn reactions. Before the eyes open, the vibrations of the parent landing on the nest cause them to gape vertically, as mentioned in an earlier chapter. When the eyes have opened, the gape is directed at the parent, or at any moving body remotely resembling a parent. From this point, presumably, they become conditioned to taking food from the parent alone. They are conditioned to the sight of the parent, so when they are orphaned, and everything is strange, including the instrument holding the food, they fail at first to respond. They must be taught, and in this teaching it is important always to use the same forceps, the same food bowl and, for that matter, the same person. Once they have relearned the art of feeding, anybody can feed them.

It would appear that the education of young birds, certainly of these two magpies, is assisted by their tendency to copy the actions of each other. Thus, when the magpies had learned once more to gape, as soon as one opened its beak the other would do so also. Should it happen that they were sitting quietly on their perch side by side, as was their habit, and later one of them moved, the other would turn and open its beak towards the first. At once, the first one would then open its beak also, and on many occasions we saw them perched side by side, opening their beaks at each other and squawking in the usual supplicating manner.

The same copying was seen when they first started to preen themselves. One initiated the action and it was followed almost at once by the other. Moreover, they imitated the details of the preening. If one fluffed the breast feathers and probed among them with its beak, the other did likewise. When one decided to preen its left wing, the other did the same. Later, this became more pronounced. Even when the two were on perches at opposite ends of the cage, so that they were four feet apart, if one raised itself on its long legs to stretch, or spread one wing downwards and stretched a leg at the same time, the other would follow with less than a second's interval.

It would be wrong to say that they copied each other always, minutely and in every detail, but they did so more often than not. So it was possible to see a general cohesion in their actions. If, once they were feeding themselves, one flew down from a perch to feed, the other would follow. When one flew from one perch to another, the other would follow. It was easy to see that if all young birds have this same habit of copying the actions of another, the task of the parents must be materially assisted.

It was not easy, however, to sort out what the family of sparrows, consisting of seven individuals, were doing, especially as they formed part of a congregation of other birds assembled on the lawn, and were mingling with them. At least we could see that they all flew down together and flew off together, it being enough for one of them to initiate the movement for the rest to follow. It may even be that because there were only two magpies the tendency to copy each other was accentuated. From them, however, it was easy to see how a combination of individual actions and a tendency to copy the movements of others can allow for individual freedom yet ensure the necessary cohesion to keep the family together.

Imitation or mimicry of bodily actions is most pronounced in the starling family and persists more strongly into adult life than in most other birds. A simple example can readily be seen when we approach a group of starlings perched in a shrub. They all fly off simultaneously, as if at a word of command. If we watch carefully, however, we see that one of the group, the first to be alarmed at our approach, makes the intention movements of taking off: it flexes its legs and half-opens its wings. Should we then 'freeze', so that no further cause for alarm is communicated to the bird, the intention movements (intention to become airborne) subside. If we did not freeze, however, the starling would take off. Meanwhile the rest of the group will have noted the intention movements of the one and imitated them, so that when it takes off all follow suit. The result is that all fly off with so little time bet-

ween each that the actions of the whole group appear simultaneous, as if at a word of command.

A more spectacular effect of this same process can occasionally
be seen in a flock of starlings flying overhead. The outstanding
occasion in my own experience occurred some years ago. I was
with a colleague, professionally a palaeontologist but a bird-
watcher by hobby, when a flock of about three hundred starlings
passed overhead, at no great height. They were in perfect formation, evenly spaced in an almost perfect oblong. Suddenly, as
they arrived over our heads, every individual in the flock turned
left and, keeping perfect formation, the oblong still intact, the
whole flock turned at right angles to its original line of flight. The
effect was almost breath-taking and the two of us gazed upwards
spellbound as the flock repeated this precision movement again
and again.

Finally my colleague broke our silence.

'Did you see that?' he asked. 'They all turned at precisely the
same moment, as if at a word of command.'

We then fell to discussing it. At that time, in the late 1920s,
there was a good deal of talk about 'the group mind'. The idea of
intention movements had not been hit upon or if it had it had
not percolated through to the rank-and-file scientists. So 'the
group mind' it had to be for us since the only alternative explanation for so spectacular a demonstration as we had just witnessed
must lie in one of the flock giving a word of command. And that
was inconceivable.

Such perfection in manoeuvre is seen only rarely, even in starlings. So is another piece of concerted action, which must have
its roots in this highly pronounced facility of adult starlings to
imitate or mimic their fellows. Here, in the grounds of Weston
House, the starlings forage all day. As the sun goes down they
take themselves off to the nearby city of Guildford, there to roost
on the windowsills and ledges of the main buildings. Usually we
see them go in twos or threes or in small groups, always departing along the same compass bearing. Sometimes they will gather
into medium-sized flocks in the tops of the taller trees, especially
of the tall lime already mentioned, and then, in small groups
going separately or in a large group simultaneously, they take off
and fly in the direction of Guildford. There is nothing spectacular
in any of this. On three occasions only, in the twenty years we
have been here, has something more spectacular occurred. It may
have happened more often and I failed to be in a position to

notice it, but I cannot claim to have seen it more often.

This particular spectacle starts off with a few starlings gathering in the topmost twigs of a very tall tree. Other starlings fly in, in ones, twos or threes, from all directions, to join them and add their voices to the mounting chorus, until a quite unusual babel of sound fills the air. Then, suddenly, all sound ceases, as if it has been cut off with a knife, or at the urgent fall of a conductor's baton. There is a complete silence. A moment or two later, the whole flock takes wing, heading in the direction of Guildford. But, having flown ten or twenty yards, the last half-dozen, as if in complete accord and as if knowing exactly what is needed, turn about and fly back to settle once more in the tree-top, leaving the rest to disappear into the evening haze over Guildford.

Having returned to their perches, the half-dozen resume the chorus, on a small scale. Other starlings, in ones, twos and threes, fly in from all around, adding their voices so that the chorus builds up once more. Then suddenly, when the symphony is at its height, the chorus ceases, again as if cut off by a knife. Again there is a short interval of dead silence and again the assembled starlings take off in a body with a whirring of wings. Once more, the last half-dozen return after having gone so far, and again the chorus starts to build up only to cease as if at a knife-stroke. After this sequence has been repeated three or four times, the returning half-dozen are joined by only a few more, a mere handful of late-comers. Finally these few leave for Guildford and this incredible orchestration is over for the day.

Striking as this abrupt termination of the chorus is, there is no need to look for a group mind or a word of command to account for it. It could all spring from a unanimity of action triggered by the intention movements of the first individual that starts to take off for the flight to the roosting area. As that individual flexes its legs and half-spreads its wings it ceases to sing, the action being copied by all the rest more or less simultaneously. The only marvel in it is the speed with which the intention movements are copied by all members of the assembled flock.

One of the more interesting features in this pattern of behaviour is the return of the last few of the flock, seemingly to act as a rallying point for the late-comers. That, however, is not peculiar to starlings. I have seen it to a marked degree in gulls setting off for a roosting area from their daytime foraging grounds.

14 *Sally, the donkey*

It was while I was on holiday in the Isle of Wight, many years ago, that I found myself driving along the road bordering an area of common land known locally as The Green. The main road, if such it could be called, ran past one end of The Green. From it a gravelled track turned off at right angles, ran up a steep hill for about half a mile, wound round a group of houses and then lost itself beyond in a cart track that ended in some fields. Not the best route for a novice driver in a second-hand car – especially the steep hill.

On its way up this hill the track was bordered on each side by greensward wide enough to give rough grazing for a few animals. On it two donkeys could usually be seen. One of these, I was later told, had the disagreeable habit of making its way to the road, as soon as a car turned the corner, and walking slowly in front of it all the way up the hill.

At all events, as soon as I turned off the main road the donkey, which had been standing doing nothing – the main preoccupation of donkeys – looked towards the car. Then it nonchalantly made its way towards the gravelled track and, timing its arrival with remarkable precision, placed itself in front of my car and proceeded to amble up the hill, keeping all the time in the middle of the track. There was no chance of passing it because there was a ditch on each side of the track, dug, I suspect, by someone in league with the donkey! There was nothing for it but to increase speed and bump into the ass's hindquarters with the offside wing. Otherwise I would never have got up that hill.

It would be difficult to say, with any degree of accuracy, what went on within that thick skull. Animal behaviourists would doubtless argue that it was a conditioned reflex that prompted the donkey to behave in this way. The argument would probably run that the donkey was used to being harnessed in front of a cart, that it mistook my car for just another cart and tried to take its normal place in front of it.

Against this can be put the observed fact that no matter what the speed of any approaching car was, this donkey would set off at such an angle that it would arrive on the track just in time to get in front of the car. I know because I went that way several times during the next two weeks. If you increased the speed of the car the donkey quickened its pace. Always it managed to get on the track in front of you and make you reduce speed to its pace – a pace that was just about sufficient to prevent the donkey, and you, from going backwards. There was something maliciously mathematical about the interception.

If this sounds a little far-fetched, I would recall the account I read some years ago, in an eighteenth-century book. I have forgotten the title of the book but remember the description of how donkeys descend precipitous tracks of the Alps or the Andes. When a donkey comes to a steep drop, with the mountain on one side and the precipice on the other, it stops and appears to be mentally ruminating. At such moments nothing will make it take another step until, presumably, it has made up its own mind. Then it places its front hoofs firmly on the track, as if determined not to let them go forward, brings its hind hoofs forwards and lowers its hindquarters, as if about to lie down. It then slides down the steep slope in a sort of toboganning, accurately following every twist and turn in the track. If this is anywhere near the truth, it is sufficient to give the lie to the alleged, and traditional, stupidity of the ass.

I recalled all these things when Sally came to live with us. The manner of her coming was, to say the least, unorthodox. We were about to sit down to supper, soon after we came to live in Albury, when my brother-in-law, who was living with us then, came in. He had called in at the Drummond Arms, the local inn, and there, it appears, he had encountered someone of the name of Mattie Williams. Mattie, so far as we could gather, had been drinking a little freely, and was prepared to tell anyone who was disposed to listen to him, that he had left a donkey outside, the oldest donkey in England. This donkey, so he said, was destined for the knacker's yard unless Mattie could find somebody to buy her for twenty pounds.

The village of Albury at that time consisted of little more than a narrow winding street with houses on each side and without any form of street lighting. Only occasional cars came through after dark, in those days, but such cars as did come by ignored the speed limit signs, so anything could happen to a donkey wandering aimlessly along the street.

Having but recently bought Weston House, and laid out large sums, far beyond my means, in having it made habitable, impecuniosity was once more the order of the day. The price being asked for Sally was exorbitant. But somehow the ready cash was scraped together and a short time afterwards we heard the clatter of donkey's hoofs coming through the entrance gates. The decision to acquire Sally had had to be instantaneous. The motives behind the decision were largely humanitarian but, as always, I had the lurking desire to add to our menagerie. Both were enthusiastically supported by Jane, who only under extreme circumstances turned down the opportunity to add one more animate being, whether furred, feathered or scaly, to our collection.

Once Sally had been installed, information about her past history began to filter through to us, in the way that information always filters through in a village. She had, it seemed, belonged to Tinker Smith. Every year for some time past, she had been entered for the Donkey Derby at Godalming, seven miles from Albury. Tinker Smith, so it was said, always appeared with Sally at the Donkey Derby, dressed in a grey morning suit with a grey top hat, the traditional dress for more elevated occasions.

There was one year, we were told, when the judges at the Donkey Derby refused to allow Sally in the race, because they said, she was in foal. 'She can't be,' exclaimed Tinker Smith indignantly. 'That's quite impossible.' Nevertheless, so the story goes, she did subsequently give birth, and then her owner remembered that one night he had pastured her in a field with a Shetland pony!

If the story is true, I would dearly have liked to have seen her offspring.

Once Sally was ours, Parkinson's Law came into operation, as indeed it is apt to do with every new inmate to a sort of shoestring menagerie like ours. First, there was the harness. Tinker Smith had her set of harness – at a price. Then we needed a tethering iron and chain, a pail for her food and somewhere to keep the harness. For the last of these a dim passage in the recesses at the back of the house was fitted with the necessary brackets where the harness could be hung, and this was named, with a flourish, the harness room. We already had a sort of loose-box beside the coach house but that had to be re-conditioned and given a new concrete floor.

Jane, at that time, was an enthusiast with a ciné-camera. She planned to make a film sequence of Sally and nothing would suffice but to buy an ancient phaeton from a nearby antique dealer.

The only time it was used, if I remember correctly, was when Jane harnessed Sally to it and drove it up the hill to Albury Heath when the Annual Flower and Vegetable Show was being held. She drove it up and down the length of the cricket field, on the heath, loaded with children at a penny a ride, to raise funds for the local Produce Association.

The film sequence never saw the light of day!

Early in the morning of the day following Sally's arrival, Jane could hardly wait to put the halter on the donkey's head and lead her round the garden, a tour that totalled the best part of a quarter of a mile. She explained that she wanted to make the donkey acquainted with her surroundings so that she should quickly feel at home. But my guess was that an additional factor would have been the sheer joy for Jane at having one more animal to handle.

The two set off at a pace dictated by Sally, which meant that she barely put one hoof in front of another. That and the doleful expression on the donkey's face, her general air of weariness and patient resignation, combined with the knowledge that Tinker Smith claimed he had had her for forty years, made Jane take pity on her and brought the thought to her mind: 'Poor old thing. We shall have to be patient with her.' When they had completed about a third of the distance, however, with Sally appearing at every step to come almost to a halt from sheer inanition, Jane saw a small stick lying at the side of the path. She stooped and picked it up.

There was no need to use the stick. The effect was electric. Sally immediately became transformed. She brightened up, her head rose and she fairly trotted the rest of the way. Unfortunately, it did not occur to Jane to throw the stick away, to see whether Sally would revert to her previous snail-crawl. Donkeys are like that. Enduring, almost tireless, capable of carrying heavy loads over rough ground, yet never putting themselves out physically more than is necessary.

There is a story of a string of donkeys used for carrying salt, in panniers. At one point on their daily journeys they had to cross a stream. One of the donkeys had somehow discovered that if she lay down in the stream for a short while her load grew lighter. They cured her of this by filling her panniers with cement the next time!

When Sally had cropped the extensive lawns of Weston House, it was necessary to rent an adjacent field to pasture her. Before long we had bought another donkey, also said to be on her way to the slaughter-house, to keep Sally company. Out of all this there was, however, an important gain. I now know the answer to the behaviour of the donkey that used to graze on The Green and make herself a nuisance to motorists. Sally was a gentle soul, as befitted her (alleged) old age. Remembering the tricks she played on us I am fully convinced that the donkey that walked in front of motor-cars did so from a sense of fun. And recalling the alleged cleverness of donkeys carrying burdens through the passes of the Alps and the Andes, I am convinced that there is no other animal that combines such a moronic expression with so much diabolic cunning as a donkey.

Sally's companion, much younger and less dignified, was promptly named by Jane: she called her Miniheehaw, or Minnie for short. There was something most fitting about giving her an absurd name. Her bray was absurd. It was more like the giggling of a silly girl. When the first few notes started up, even after we ought to have been used to them, we had to stop and think whatever this strange noise could be, until we remembered it was Minnie.

I had always supposed that the braying of an ass was a stereotyped sound, much the same in all donkeys. This, at least, was not true for our two. Sally would start off like a foghorn and the heehaw part of the bray was delivered in almost ultrasonic frequencies, so high-pitched that you could hardly hear them, although you were well aware from seeing Sally's head up, upper lip turned up and mouth open, that she was making sounds of some kind. And there was a plaintive quality about her bray.

Donkeys are for the most part silent animals, which is responsible, one would guess, for their reputation for patience in adversity. Sally never brayed for the fun of it. She used it when, tethered, she had eaten all the grass she could reach and wanted to be moved to a new patch. We did not appreciate this at first, but once we had got the idea we recognized the special quality of the bray which indicated that she wanted to be moved. She also brayed if her morning mash was a few minutes late. When she wished to go into the loose-box she brayed repeatedly and at regular intervals. We quickly found we could interpret the meaning of her calls: the braying was a language and she soon taught us the meaning of it.

Every so often somebody comes up with a bright idea, to take an animal into the laboratory and, with the aid of up-to-date apparatus and painstaking research, study the sounds it makes.

The scientist discovers they constitute a form of language. Tinker Smith, living close to Sally for many years, reached the same conclusion using no more than his ears and his native commonsense. We were soon aware of it, too. Of course, the laboratory method saves time and gives more precise results – but the rule-of-thumb method used by the tinker and by us is more fun.

When, after Sally had been with us for a year, the tinker called to see her again, we were left in no doubt that the donkey was not wholly ignorant of the meaning of words. Apparently she had been given a glass of beer from time to time. 'Watch this,' said Tinker Smith. Then he said, 'Who wants a glass of beer?' Without more ado Sally pushed him in the middle of the back with her face, carrying him a few steps forward. This, I was given to understand, was her way of pushing him through the door of a public house. No doubt he had trained her to do it, so there was nothing greatly remarkable in it, any more than teaching a dog tricks, except that one does not expect it of a donkey. Another reason why one did not expect it was that Sally's face was so lacking in animation. She always looked doleful; her eyes lacklustre, she looked as if she would fall asleep at any minute. This did not mean the brain was not working.

If Sally was on the ground when one of us went to her with a halter in our hand she immediately laid her head on the ground and closed her eyes completely, as if fast asleep. It then became necessary to lift her heavy head to slip the halter over it. If you fumbled, so that you lost your grip on her head, it straightway sank to the ground as if Sally were truly in a deep slumber. She never actively resisted having the halter put on her head but if she could put difficulties in the way without too much effort on her part she would do so.

There came a time when it was necessary to have both the donkeys shod. There was a blacksmith's forge in the village then and the blacksmith obligingly came to us to do the work. Minnie was the first to be attended to. She had probably not been shod before or if so not very often. The farrier sweated and panted in his effort to keep control of the young donkey who all the time was trying to dance a jig. Fortunately, the farrier had a lively sense of humour and the whole proceedings were accompanied by a mixture of his laughter and the clatter of Minnie's hoofs.

When it came to Sally's turn she walked over to the farrier, as quiet as a lamb, lay down on the ground and closed her eyes. So far as we could see she slept peacefully throughout the whole proceedings. The blacksmith's comment on this was: 'All sorts of things have happened to me but I have never before had a patient go to sleep on me!'

Ten years before we had Sally, I was being driven through the New Forest by one of the verderers. One of the main responsibilities of a verderer of this ancient forest, known to all schoolboys as the place where William Rufus, a son of the Norman Conqueror, met his death, is to see that the laws of the forest are not broken. He started to tell me that, by ancient statute, the ponies and donkeys in the New Forest have the right of way on all the roads. Just then three donkeys came into view. They were walking in single file on the crown of the road and, as is usual with our asinine friends, they were not in any hurry. This alone made two-way traffic difficult, but when the middle donkey leisurely prostrated itself and commenced a luxurious and unhurried roll in the dust, while the other two stood and looked on, all traffic was at a standstill until the mokes chose to move on. It is at such times that one realizes just how long a donkey can spend on this exercise, and how much it seems to enjoy it.

Sally did her share of rolling in the dust but, what was more surprising, she enjoyed nothing so much as rolling in hot ashes. After a garden bonfire had burned itself out, and while the ashes were too hot to handle comfortably, she would, given the opportunity, lie down in them and roll in seeming ecstasy for longer than she ever rolled in dust. It is not surprising that mankind has made so much use of fire and the heat from it when we see how the four-footed beasts seem to enjoy it. It is a common sight, or used to be when hedges were more numerous, to see cows congregating around the remains of a fire where hedge-trimmings are being burned. After the fire has died down one or more of them will stand over the hot ashes. This cannot be wholly explained by saying that in temperate climates there are days without sunshine, even in summer, and that animals find heat from fires a good substitute.

Dogs, we are told, can readily endure low temperatures. They also seem able not only to tolerate great heat but actually to be drawn towards it. Our boxer-cross, Jason, used to demonstrate this. When I had been burning garden rubbish and the heap was blazing well I would be unable to approach within six feet of it. The dog, by contrast, would go and lie down three feet from the edge of the burning pile. He was clearly uncomfortable – that could be seen from the way he turned his head first one way, then the other, and twisted and turned his body – yet he persisted in staying close to the source of heat.

So far as smoke is concerned there seems to be an even more inexplicable situation. Again and again, during the twenty years we have lived at Weston House, when the garden rubbish is being burned in a corner of the garden farthest from the house, the cows in the surrounding fields have walked deliberately over to the billowing smoke. They cannot get nearer than twelve feet of the fire, at a distance where the heat must be barely perceptible, but the whole herd will gather and stay for a period of time, apparently inhaling the smoke.

It may be quite true that the human habit of smoking is to be deplored as detrimental to health. Yet the fact remains that mammals and birds seem readily to become addicted to smoke, including chimpanzees in zoos that are given access to lighted cigarettes.

In *Animals* for September 1967, Mr E. N. G. Earle, of Brecon, South Wales, described how his terrier, then approaching his fifth year, had all his life been interested in smoking and smoke in general.

> He is delighted if we pick up a box of matches as he knows something of interest is about to take place. If we light the fire he will watch the smoke going up the chimney with joyful barks; he will then demand to be let out of doors, where he will look up at the chimney and bark away and run frantically up and down the lawn watching the smoke come out.
>
> He is delighted by cigarettes, cigars, or pipes, and will jump over two feet to catch the smoke in his mouth. He will look at cigar- or cigarette-ends on the ground, and will jab at them with his mouth if they are still smoking, eventually picking them up and taking them away.
>
> He is enthralled by bonfires, and unless watched carefully will get right into them. If there is a good breeze he will chase the smoke as far as it goes and will stay all day with it if he can. All of this is entirely 'self-taught' – I have never consciously taught him any tricks.

Since then I have collected a number of similar instances, even one of a toad that swallowed cigarette butts: it regularly visited a verandah in India, where British officers played bridge in the evenings, to pick up their butts.

The wild asses of south-western Asia and north-east Africa are typically desert animals, able to live on coarse vegetation, largely independent of water, although, like the camel, drinking copiously when it is available, and having greater endurance and hardiness than the horse. So the donkey, easy to maintain and to feed, domesticated descendant of wild asses, became the poor man's beast of burden. If donkeys are ill-treated it is not from an inability to defend themselves, however. In my boyhood days I was on one occasion witness to the power of a donkey's hindlegs. I don't know what had upset the animal but I do recall seeing it lash out with its hind hoofs and shatter the door of its stable beyond repair. Thomas Pennant, the English zoologist, tells of seeing a fight between a donkey and a dog in the bear garden in Paris, in the eighteenth century. The dog was unable to get a grip on the donkey, but the donkey sometimes held the dog in its teeth and sometimes flung it under his knees and knelt on it. It was the dog that finally gave up the fight.

Moreover, any donkey that does not elicit sympathy from its human masters is unlucky for it starts with a natural advantage. Its long face, expressionless eyes and drooping lower lip impart a doleful appearance to the face and give all but the most sprightly donkey an air of resigned patience.

Sally exploited this natural advantage to the full when she and Minnie were turned into the field; our lawns could no longer feed them adequately. Both were brought into the loose-box for the night and taken out to pasture in the morning. The field was the other side of the street from the house and soon the two asses became the object of a daily pilgrimage from the village. Mothers brought their children to see the donkeys and to feed them, usually with bread.

Iron railings separated the field from the highway. Minnie accepted food when encouraged to come to the fence to take it. Sally, in the maturity of her years, knew better how to play on the feelings of her visitors. She pushed her head through the railings so that it looked as though she was trapped. And there her long face was exposed for all to see, with its lugubrious expression calculated to wring the withers of any but the most hardhearted. The first time Jane saw Sally with her head through the railings she supposed the donkey had accidentally got herself trapped and could not free herself. Jane manipulated the heavy head, Sally doing nothing to help her. At last, with much heavy labour Jane managed to turn the animal's head and withdraw it. Sally, released, merely stood there, with no change of expression on her face. Then when Jane had walked away, she took a step forward and pushed her head through the fence, as if to say: 'This is how I arouse their sympathy and get the titbits.'

It must be confessed that in spite of all that is said here there was never anything remarkably striking about the behaviour of our donkeys, even of Sally, the sage old matron. We never saw more than I've recorded here to justify crediting donkeys with special wisdom. That is the way things go. The special acts of cleverness are seen by accident and occur but infrequently, like the donkey that, when tormented by flies, would pick up a sack in its teeth, turn its head sharply and flick the sack over its back to disturb the flies – and was photographed in the act.

There is also the account of a string of donkeys being used for transporting rocks, in panniers, from one place to another, for road-making. One of these donkeys had a high IQ, evidently. When the donkeys were returning to be loaded, this one hung behind and was as a consequence the last to have its panniers filled. When they were led back to the point where the panniers were to be unloaded, this same donkey contrived to be at the head of the column, so it bore its loads of rock for appreciably shorter periods of time than any of the others!

There is always danger in generalizing from a single instance and perhaps I tended too readily to see Sally as the epitome of all donkeys. If the truth be told there was a vast difference between her and her companion Minnie, who was, we were told, only two years old when she came to us.

The one person who was most aware of this difference was Wally Fry, our gardener, for it was his task to take the two don-keys into the field in the morning and lead them back to the loose-box in the evening. In order not to have to make two journeys each time, he led the two donkeys together. He is fond of describ-ing how one of them wanted to race ahead and the other was slow to start and refused to quicken her pace. So he often found himself walking from field to loose-box, or vice versa, with his arms stretched to the limit trying to hold Minnie in check while at the same time trying to pull Sally along.

One evening, at the time when the donkeys needed to be fetched from the field, thick fog descended on the valley in which Albury lies. Minnie was her usual self, eager to trot to the loose-box. Sally chose this particular evening to be at her most stub-born, and she stopped dead in the middle of the street. Had a motor-car come along at that moment there would have been car-nage. It had its benefits, however, because it prompted Mr Fry thereafter to tie a piece of rope from one to the other of the donk-eys' halters. So from then on, Minnie pulled Sally along while

Sally held Minnie back.

One morning the door of the loose-box was found to be open. Sally was lying on the floor half-asleep but there was no sign of Minnie. She had managed to open the door and was later found wandering in the village. Somehow she had lifted the latch on the outside of the lower half of the loose-box door. Sally, evidently, was completely without interest in the proceedings. She was still half-asleep when Minnie was brought back and did not bother even to lift her eyelids when her younger companion was ushered back into the loose-box. She had probably seen it all before!

Masterly moorhen chick 15

A moorhen sitting on its nest among the water plants that make up its natural habitat is aesthetically satisfying. In the spring the fresh green of the vegetation dappled with the blue of the water forget-me-not makes a pleasing setting, especially when it is bathed in sunshine. Against this the glossy slate-grey of the moorhen's head and neck and the brown of its wings contrast vividly with the sealing-wax red of the bird's frontal shield and the bright yellow at the end of the beak. Add to this the white 'Plimsoll Line' on the wing and flank and the white of the under-tail and the adult moorhen is a very handsome bird indeed. But we can rarely appreciate its beauties. This is largely because it gives us little opportunity to see it closely for long at a time, simply because it is so shy.

The newly-hatched moorhen is even more colourful. Its body is covered with black down and the legs are olive-green. At the back of the head the skin shows pink through the black down. The face is ornamented, as in the adult, with a bright red frontal shield, the bill is orange with a yellow tip and there is bare blue skin around the eyes. There are two odd things about the young moorhen. One is that the newly-hatched chick has a red bill, like the adult and that this becomes green at three weeks of age but turns red again the following winter. The other is that the newly-hatched chick has a claw on its thumb, recalling the condition found in that remarkable, primitive bird of South America, the hoatzin. The young hoatzin climbs about the nest and among the branches and twigs of the trees and shrubs in which the nest is built, using the claws on its wings as well as its feet. The baby moorhen is, for the first few days of its life, unable to stand on its own legs and progresses by scrabbling with its feet and using the long claw on each wing in the manner of a disabled person putting his weight on two sticks.

I had been interested in moorhens for a number of years when, in 1960, a pair of them nested in the Tillingbourne, the meandering stream that flows through the fields across the street from

Weston House. The stream there is not very deep and just beyond the brick bridge that carries the path into the woods a small sandbank had accumulated in midstream. The moorhens had built their nest on the highest point of the sandbank.

The foundations of the nest were made up of dead reeds, flags and portions of other water plants put down in a surprisingly neat basket-work, suggestive of a high degree of craftsmanship, even of artistry. The materials added later were less orderly and obscured the neatness of the foundations. There was also a suggestion of artistry in the way the margins of the nest were decorated with flowers and pieces of paper.

The number of eggs in a clutch is usually given as five to eleven. It may be as low as two or as high as twelve to twenty-six. It has been suggested that any clutch over fourteen is probably the work of two hens using the same nest, but I am inclined to doubt this, for a feature of moorhens is the high mortality rate among the chicks, linked with a high rate of reproduction. The hen on this particular nest had laid twelve eggs and one morning I saw Jane wading into the stream, while the parents were absent, and lifting one of the eggs, to carry it in her hand to the bank. Any law-abiding citizen is conscious that there are laws protecting birds and their eggs and that, even without such laws, it is reprehensible to interfere with nests.

I assumed Jane had a good motive for her action but that did not prevent my asking her what she was doing, one of those senseless questions that serve only as a preamble to further conversation, since it was quite obvious she was taking an egg.

'I'm taking one of the moorhen's eggs,' Jane replied. 'I'm quite sure she won't miss it and anyway most of the chicks will die from one cause or another.'

In fact, we later saw how the rest of the brood disappeared one by one until only two survived. What we also saw was the occasional tell-tale clue which indicated how great are the hazards for infant moorhens in their natural environment.

'What are you going to do with it?' I asked.

'I have a broody bantam and this seems an excellent opportunity to study the development and behaviour of a young moorhen. In any case, I shall take care not to hand-tame it and when it is able to look after itself I shall let it go, back to the wild.'

Any qualms I may have had were stilled. I had formed the impression that the moorhen is among the more intelligent of birds. This was based on a very few observations and having a moorhen about the place might yield interesting new information.

The name of this bird is variable from one part of its range to another. Explaining this gives us some preliminary information about the species. By one of those perverse tricks of common names the bird is known as the moorhen in southern England, where moors are in short supply, and as the waterhen in the north where moors are a feature of the countryside. The first name dates from the twelfth century, the second from the sixteenth. Then in the eighteenth century the species was given the generic name of *Gallinula*, from the Latin meaning a little hen, presumably from a fancied resemblance to the domestic fowl, although the two are not related. From this was derived the vernacular name gallinule by which it is known in the United States, where it is found in the south-east.

The alternative names 'moorhen' and 'waterhen' suggest an ornithological marine, at home on land or on water, and this proves to be the case. I can think especially of one community of these birds that spends the whole day searching the grass for food in meadows bounded by meandering streams no more than a foot deep and a yard across for most of the year. The birds use them only as a refuge to run to, which they will do when an intruder approaches within a hundred yards of them. Another community half a mile away feeds exclusively in the water. The truth is that 'moorhen' is from 'merehen', 'mere' meaning 'a lake'.

Moorhens will perch in trees. They swim and dive well and I have seen one fly strongly for a hundred yards at tree top height. But for the most part, whether foraging on land or escaping from a disturbance, moorhens prefer to use their feet. Taking all in all, if the egg we had secured hatched successfully and the chick was reared, we had a subject for study that would not be incommoded through living in a garden. It need not be restricted within an aviary, either, once it was full grown because it would be unlikely to seek escape in flight. Conditions for study could not be better.

When the moorhen chick was hatched the bantam foster-parent began to do her duty within the limits of her mental equipment. She did what any self-respecting hen would do for her brood. She searched for food and when she had found it she called to the chick to come and take it. The clucking of the hen found no response within the brain of the young moorhen. Had the moorhen been left with its natural mother it would have received a different treatment. That mother, or the father, would have looked for morsels of food and would then have brought them to their offspring, for the moorhens' method of feeding their chicks is to hold the food in front of the infant's beak. Not only does the chick immediately respond by taking food proffered in this way, it also goes towards the parent advancing with the food. At the same time it calls to the parents with a *peep-peep*.

Here, then, was deadlock. The bantam foster-parent did not understand the significance of the *peep-peep*, or at best only partially understood it. The moorhen chick failed fully to comprehend the hen's clucking. The bantam fosterer expected the chick to come all the way to her to pick up the food she was pointing at with her beak. The chick expected to go only halfway to the hen and to have the food held out for it to take. So it became necessary for us to hand-feed the chick by holding food in front of its beak. Surprisingly, it quickly learned to run towards this second kind of foster-parent, the gigantic human that loomed in front of it.

A lesser being than a moorhen chick might have been completely bewildered by this choice of fosterers. When one has handled scores of orphaned baby birds, as we have done, one learns that some are easier to feed than others, even those of the same species, and that there are even greater differences as between different species. This may sound like stating the obvious, but there is more to it than that. As a result of experience one soon begins to see the responses by the babies as a rough guide to their intelligence.

In this instance, the chick was faced with a foster-parent that was near to its natural parent in size and appearance but with alien ways. Alternatively, it found itself being helped by huge un-birdlike apparitions that to some extent at least understood moorhen feeding tactics. It could have been excused had it developed neuroses from perplexity or anxiety, but there was no hint of that and, without being able to say precisely why, one had the impression of a young bird behaving with greater intelligence than is usual in young birds. The only positive indication we had was that within a week the young moorhen had learned the significance of the hen's clucking and responded appropriately, while the hen, so far as we could see, never did learn to understand the chick's peeping.

Before going further with the history of this particular chick, it may be worthwhile recalling what is already known from observation of moorhens in the wild. This indicated that the young of the species is unusually precocious. For example, on 20 June 1955, in Kensington Gardens, Mr Robert Hayman, an experienced ornithologist, had watched moorhens doing something unusual. Their nest, originally sited among willow stems at the head of Long Water, had become detached, possibly as the result of a rise in the level of the water, and had drifted into the open. It had apparently become halted by a piece of plant, in shallow water, some twenty-five yards from its original site and about ten yards from the bank. The top of the nest was then about four inches above the level of the water. Moorhen nests are liable to flooding and the parents' reaction to this is to add fresh material to it to build it up. Indeed, this reaction can be said to be latent all the time, for throughout the breeding season, even when there is no danger of flooding, they add fresh material to the nest not in great quantity, but fairly regularly.

The interest in Mr Hayman's observations was that he saw a parent moorhen and two half-grown youngsters combining to build up the nest to above the danger level from further flooding. He saw the parent bring a fair-sized stick to the nest and present it to one of the young ones, which took the stick and, as the parent swam away in search of more material, worked the stick into the rim of the nest. During the next quarter of an hour, he saw the old bird bring several fair-sized sticks and leaves of water plants, all of which the same young one took one after the other and worked into the fabric of the nest. With one awkward piece of material the young bird had to get up and move around to dispose of it satisfactorily.

A similar set of observations had been made a few years previously by a German ornithologist, and it has doubtless been seen since by other observers. In addition, at about the same time as Hayman's account was published, I received a letter from Dr W. D. Lang about a scene he had witnessed. It was known already that the young from the first brood will help the parents feed the nestlings of a second brood. Similar behaviour has been noted in a few other species, such as bee-eaters and American jays. What Lang reported was that a young moorhen, attending to a nestling, was having difficulty in doing so and, apparently, swam off to fetch one of the parents to the nest to deal with the situation.

This seemingly marked precocity in the juvenile moorhen seems to be matched by the adults, if we accept the word of three anglers. They were fishing a stream when it began to rain and one of them noticed that a moorhen, sitting on a nest nearby, pulled a piece of plastic material that was lying near its nest and placed it over its back, tucking it in at the sides. When the rain stopped it removed the plastic material. Later, it started to rain again and all three men saw the moorhen cover its back once more, and apparently this was repeated several times during the course of the day.

This story is sufficiently remarkable to engender scepticism in some that have heard it recounted, but it may be that moorhens, or some moorhens, protect themselves with leaves in this way. Certainly it has long been known that a moorhen may draw down the sword-like leaves of iris to form what looks like a decorative canopy over the nest. It has always been assumed that this is done to hide the nest but it may be a protective measure against the elements. Even so, the use of the sheet of plastic material against rain is still remarkable, especially that it should have been removed once the rain had stopped. So it was against the background of this sort of information that we watched the development of our foundling chick with interest.

At the age of twelve days the chick was able to feed itself completely so it was parted from its bantam foster-mother and placed in an aviary six feet by six feet by six feet high. We had adopted the policy when we moved to Weston House of building units that could quickly be put together and taken apart again to make aviaries of almost any size. Each unit consisted of a square frame of inch-by-inch quartering with a cross-strut at the middle to give rigidity, the whole being covered with small-meshed wire-netting. Every sixth unit was fitted with a door, made on the same principle, of a frame of inch-by-inch quartering covered with wire-netting.

Our smallest aviaries were therefore of six-feet sides and height and were light enough for two people to carry easily to any point in the garden where an aviary was needed. Using eight units an aviary six feet by twelve feet by six feet high, with its door, could be quickly put together. Eleven units would give an aviary six feet by eighteen feet by six feet high, and in such an aviary even a large bird such as a crow could spread its wings and really fly instead of merely jumping from perch to perch.

When not in use an aviary could be dismantled and the units stacked in a relatively small space. They could also be used, assembled, as pens for small mammals, such as stoat, weasel or polecat, in which case a ceiling unit would be used on the floor to prevent the animal burrowing its way to freedom. Larger aviaries were also used but these were permanent, made of tall poles driven into the ground, strengthened by cross-bars, with the whole framework covered with wire-netting.

The smallest of our portable aviaries was large enough to give the young moorhen ample space and freedom of movement. We noticed that whenever one of us went into its aviary it always retreated to the farthest corner. Since the aviary was seldom visited, except to take in food and water, this alone calls for comment, expecially in view of what has to be said about the moorhen's later behaviour.

An aviary bird that is really tame will at least not retreat from its human guardian. More often it will come forward when food and water are being supplied to it. It will recognize that the visitor to its home is the bringer of fresh supplies and will be anxious to get to these, and will do so with a full sense of security. A bird that is tame only in the sense of tolerating the near presence of a human being will always retreat and seems not to be able to distinguish between a goodwill mission and one that might carry a potential danger.

There is, however, a qualification to this which is not without interest. It would seem that birds, and probably other animals as well, when not fully tame to the point of feeling secure, are highly sensitive to the limits imposed on them by containing walls, as in an aviary. Thus, contrasting the later behaviour of our moorhen with that which it showed while in the aviary, we found that it did not retreat or try to get away when it was free in the garden and somebody went along to put down its food. Instead it followed that person, though admittedly at a short distance away. The moment the food was put down, however, it came forward to eat. It was as if the knowledge that it could, if need be, make good its escape at a moment's notice made the young bird correspondingly more bold.

This gives something of a clue to one of the factors involved in this state we call tameness, which is more than merely not being afraid of the human presence. It seems also to include the acceptance of a territory and of territorial limits. So far as incarcerated birds are concerned it is as if the aviary in which they have been living is their world, the only world in which they are completely at ease: it is their property, their home. In other words, captivity is not in itself a burden if the aviary, or the pen, is so furnished that it makes a close approximation to a piece of natural habitat and if the food provided is as near a natural diet as it is possible to make it.

This is my tentative answer to those who ask, as many people do, whether it is not cruel to keep animals of wild species confined. When we have taken over pets that other people can no longer house, we feel constrained to keep them and care for them for the rest of their lives. Those that are brought to us for treatment or care, usually fledglings or young animals, or older birds that are injured, we release into the wild as soon as they seem able to look after themselves. Exceptions are made where there is the probability that they might suffer or come to an untimely end if released.

The general behaviour of an animal will soon indicate whether its surroundings are such as to obviate boredom. The condition of the plumage or the fur will soon indicate whether the correct food is being supplied. And here I would digress further to make a point which I always emphasize in reply to people who ask how to treat a wild animal that has come into their possession. It is that, in using bread as part of their diet, white bread should be avoided. We always use wholemeal bread and this is not a fetish: we have found as a matter of practical experience that white bread is detrimental to their health.

Another essential to their comfort is, I think, for them to accept the aviary or pen as their territory, their exclusive property. This seemed to be illustrated by the behaviour of our genet, as we saw in chapter three. Several times we have had the experience of a bird or other animal having been released, or having escaped, continuing to live in the vicinity of its former home. Birds, for instance, will spend most of their time by day on the top of the aviary they formerly inhabited, and will use it as a roost. They will do so even when free to wander. The behaviour of our moorhen seemed to endorse this for, so long as it was in the garden, whenever it was uneasy during the day it would make for its aviary and stand on top of it. It also roosted there.

Giving the moorhen its freedom consisted of taking it from the aviary which was sited nearly two hundred yards from the front gate. It was then carried out through the gate, across the road to the Tillingbourne, its natal stream, then taken nearly a hundred yards along the bank and there released. It swam across the stream and disappeared into the undergrowth on the opposite bank. Within three days of its release it was back in the garden on the spot where its aviary, now dismantled, had stood. It may have returned earlier for all we knew to the contrary, but three days was the maximum time it stayed away, and it is worth going more deeply into this matter.

Moorhens, as I have said, fly very little. As a rule, the most one sees them do is to take wing if suddenly disturbed, when feeding away from water, and fly in a short, low, seemingly laboured flight to the nearest water. Since the surface of the water of the Tillingbourne is below the level of the roadway, across which we carried the moorhen to release it, and is screened from it by fairly continuous bushes and shrubs, it seems fair to say that from a moorhen's viewpoint the roadway and my house beyond would have been screened from it by a rising bank topped with a screen of vegetation. The garden surrounding the house itself is sur-rounded by a wall, in places topped with close-set iron railings, the barrier it presents being not less than seven feet high at any point. All round the perimeter of the garden, within the wall, to a depth varying from fifteen to twenty feet, is a more or less con-tinuous growth of shrub and tree forming a dense screen up to a hundred feet high most of the way.

To re-enter the garden by the route along which it was taken out, the moorhen would have had to negotiate this dense barrier of foliage, easy enough for a small bird or one habitually given to travelling on the wing. The moorhen may conceivably have come back through the gateway; it would then have had the house on its left followed by a high brick wall, continued as a high fence screened by dense bushes and trees, with only a small entrance through it and one habitually kept closed by a wooden gate over which the moorhen would have had to fly. In short, there was no easy or obvious return route for the bird to follow. So, in order to return it had to surmount many obstacles, and also it would have had to recognize its former home looking very different from the outside as compared with the view that it had had from inside the garden.

It may not be stressing the point too much to suggest that it must have made determined efforts to find its way back to its old home. It must have experienced a strong urge to return since, by going in any other direction from the point it was released, it had the alluring prospect of open country in which to find a place to settle down.

More striking still, once back, if anything occurred to make it uneasy it made straight for a patch of grass, six feet by six feet, on the main lawn and there it would remain until the disturbance had ceased. This small square of grass was where the aviary formerly stood.

The moorhen's return was the more remarkable in that it was only partial, for although it had come back to spend its days with us it went back to the stream at night to roost. As evening came it would fly onto the top of our eight-foot boundary wall and there it would pause. There is a fair amount of traffic along the road at the foot of the wall and, believe it or not, the moorhen would look right, look left and look right again and only if there was no vehicle in sight would it cross the road in a short, typically lumbering flight, to land just inside the meadow beyond. After that it walked across the meadow and disappeared down the slope beyond, towards the Tillingbourne.

I watched the bird go on a number of occasions and every time saw it look right, look left and look right again before fluttering down to the meadow beyond the narrow roadway. If, having made its inspection, a vehicle came into sight, the moorhen would wait for it to pass before once again looking right, left and right before flying across.

Eventually the moorhen failed to return and our hope was that it had made friends with others of its kind, and perhaps had even found a mate.

I can recall one wild animal only that purposively looked both ways before crossing a road. This was a fox. As a line of vehicles accumulated at a crossroads, with traffic lights, the head of a fox appeared in the gap between the rear bumper of one car and the front bumper of the next car. The fox looked left, looked right and looked left again before trotting across. But then foxes habitually cross roads and they have good brains. To find a bird normally living on or near water, and a bird that had previously had little experience of roads and traffic, doing so is little short of incredible.

The story has been told of a lily pond in a garden that had been emptied to be cleaned. During May a moorhen built its nest at the usual spot, under an overhanging bush, on what was then dry land. As the pond was being refilled the nest rose with the rising water and floated insecurely. The moorhen pushed the nest with her breast across the pond to a partially sunk punt, a distance of fifty yards, using a following wind. She held the nest firmly against the punt, on the lee side. Then she transferred the fabric of it twig by twig into the punt, not rebuilding as she did so, presumably because she appreciated that this would have risked the nest floating away. When all the material had been safely transferred she boarded the punt and re-built the nest. This took an hour and the next day she had laid a clutch of eight eggs.

Having had experience of our moorhen chick, with opportunities to observe it closely, I find this story, and others quoted here, acceptable. One of the most valuable results of having wild animals under close observation, for long periods of time, is that one obtains a yardstick for measuring the value of other people's observations. It is possible also to invoke my 'Principle of Non-coincidence'. If a number of remarkable observations are recorded concerning one species of animal by people separated in time and space there must be more than coincidence involved.

To add one last story to the moorhen saga we have the testimony of Sydney R. Davidson, of Perthshire, Scotland. He recorded how in a very cold spell in winter moorhens moved up from the burnside to feed at his bird tables. He scattered maize meal, birdseed and brown breadcrumbs on the lawn for them. The cold grew intense but around midday the sun melted the thin crust of snow a little, and there was a film of moisture on the ice. Then it froze again and only sparrows and finches could detach particles of food from the frozen ground. Previously the moorhens had moved about all the time they were feeding. Now, each 'brooded' for several minutes at a time, rising and eating from where it had crouched, then walked a few paces and brooded again. Mr Davidson reported that this pattern of feeding continued while severe conditions lasted. He concluded the moorhens were using their own body heat to thaw out the food.

Incredible! But believable from our own experience with our 'adopted' moorhen.

16 *Perce, the crocodile*

Soon after we came to live at Weston House we learned of a woman in Worplesdon who kept crocodiles and alligators. She had, it seems, eleven of them which she kept in her bath. Worplesdon is a village several miles away, somewhat off the beaten track. The story about this unusual lady came to us in the way most information percolates through a rural area, via that intangible, invisible channel of communication usually described as the grape vine. We paid little attention to this particular story, and especially to the part of it which told that this good lady was in the habit of taking her crocodiles out onto the common, to a pond, where the children used to play with them.

Then Perce turned up, at a time when the skins of crocodiles and alligators were in great demand for handbags and other fancy goods. 'There's not enough of him to make a handbag, only a purse,' said Jane and our poor little reptile was promptly named Perce, short for Percival. Perce was a Nile crocodile, only a baby when we first clapped eyes on him. He arrived at the house in the hands of a man who had just come back after a spell of duty in Africa. Having brought the infant crocodile all that way our visitor was now experiencing difficulties in housing it. Some form of larger grape vine had evidently put him on our trail and in no time at all Perce had been installed and his previous owner had departed.

Looking back now, the details are lost in a mist of uncertainty. It is usually so with the animals that have come into our care. Somebody arrives on the doorstep holding an injured bird or a nestling, or whatever it may be, in cupped hands. Your whole attention is on the plight of the animal, and before you know where you are the person that brought it has departed leaving no name or address.

Anyway, Perce arrived, and at that time there was a sudden vogue for keeping unusual pets, particularly crocodilians. This was around 1960. A press report of about that date told of a 'Safari' that had been planned in New York to rid the sewers of alligators that were terrorizing the sanitation workers. The report alleged that in New York alligators had become popular pets for children – so long as the reptiles were in the first stages of infancy, as Perce was. That is, about eight inches long. When they started to grow, so the story continued parents would 'sneak them away to the bathroom and get rid of them'. Once in the sewers the alligators, we were told, thrived on rats and one was said to have reached a length of nine feet. If this last claim were to have been accepted it would have meant that this method of disposing of unusual pets had been started at least ten years before, since the American alligator grows at the rate of just under a foot a year for the first ten years and towards the end of that time slows down to a mere few inches. There is a similar growth rate for the Nile crocodile.

Whether by coincidence or not, C. T. Astley Maberly, the South African naturalist, found a similar tendency towards the keeping of baby crocodiles as pets in Africa, eleven years prior to the New York press report. He himself hand-reared one and had it in captivity from 1941 to 1947. His account of this, published in 1949, contains a very complete description of the behaviour and development of the Nile crocodile. I did not come across Maberly's notes until we had already had Perce for a while. It was interesting to see what he had had to say compared with what we had observed in our own baby reptile.

When Perce first came to us his body was more rounded and the head less flattened than in the adult, and although the proportions of head to body to tail were similar to those of a full-grown crocodile, there was something puppy-like about his appearance. It was easy to see why parents in the United States had fallen for baby alligators as pets for their juniors. Also, the eyes were larger in proportion to the head than in the adult, suggestive almost of a kitten. Moreover, as in the adult crocodile the pupil contracted to a vertical slit in strong daylight and the changes in the pupil, from slit in strong light to rounded in dim light served to enhance the slight resemblance to a kitten's head.

There were practical limitations to the amount of space we could give our reptile. He was placed in a glass tank thirty-six inches long by fifteen inches across and the same deep, with six inches of water. He had room to move about and he was taken out of the tank from time to time so that he could 'stretch his legs'. The floor of the tank was covered with sand and pebbles giving various depths, so that he could rest with the top of his head breaking surface and his toes touching bottom or he could stay just submerged, suspended by his own buoyancy, with little more than his eyes and nostrils above water; or he could submerge completely. An island of pebbles at one end of the tank allowed him to haul out and bask when he felt so disposed.

We were slightly worried about temperature. Crocodiles, we had ascertained, were most comfortable and their bodily functions at their best between 75 and 85 degrees Fahrenheit. Below 75 degrees they become progressively more torpid and temperatures of 39 degrees Fahrenheit or lower are said to be lethal – which explains why crocodilians are confined to the warmer parts of the world.

The man who brought Perce to us assured us that ordinary room temperatures would be sufficient. But what is room temperature? The room in which Perce was kept was heated by radiators in winter. Also, it caught the sunshine for more than half the day and, because it was next to the boiler room, with the boiler active all the year round, the temperature seldom dropped lower than 65 degrees. In fact, it is probably the warmest room in this big old draughty house. Perce was probably reasonably comfortable, but only just, which may account partially for the times when he would go for long periods without feeding.

Maberly tells how, one August night, when a strong north wind brought a heavy frost, the water in which his baby crocodile lived became coated with half-formed ice. He found the little crocodile floating belly uppermost, stiff and apparently dead. He wrapped it in cottonwool, put it in a cardboard box and placed this on the kitchen stove. Within a quarter of an hour the reptile had revived and lived for many years after.

As a pet Perce was a novelty to us, but a somewhat dull one. He spent his time in the vivarium, either in water or out of it. He fed so seldom it made no great demands on one's time. Indeed, his ambition seemed to be to spend as much time as possible in a state of immobility and quiescence.

From time to time he was taken from his vivarium into the garden, on the assumption that he ought to have some exercise occasionally. It was then one could appreciate that crocodiles are not always sluggish. He would walk easily enough on land with the body carried well off the ground and legs stretched full length, the head parallel with the ground and tail dragging. Then he looked what he was, a survivor from the Age of Reptiles, one could almost have said a small living dinosaur. This is no more than should be expected since crocodilians have remained virtually unchanged for the last one hundred and eighty million years.

He could walk fairly fast, also, but only for a short distance, with frequent pauses to sink motionless to the ground as if summoning the energy for the next stage of his journey. This, no doubt, is why crocodiles and alligators seldom move far from water and escape back to it as soon as possible. Once in water their progress is easy and seemingly effortless, in marked contrast to their movement on land.

In water the legs appear to be used mainly as balancers and to assist buoyancy when the animal is motionless, especially the hindlegs, which are then spread. In swimming, also, the legs play a subsidiary role. The forelegs are pressed back against the body and hindlegs, although the hindfeet are webbed, are carried straight back, parallel with the tail and slightly below it. Propulsion through water is by lateral movement of the strong tail.

It has often been said that crocodiles use a whipping action of the tail to kill prey. This, it now seems, is not so, except in rare instances. The action of the tail is purely defensive. In a large crocodile the whipping of the tail may maim or kill whatever it strikes but this is incidental and a tribute more to the strength of the tail than an indication of motive on the part of the reptile. Primarily it is defensive and it may, in fact, be no more than the result of a reflex in a naturally aquatic animal, using its tail for propulsion, since in water the first movement in seeking to escape would be to move the tail in this way to propel itself through water.

Perce was only an infant and had been accustomed to being handled from an early age, which may be why he did not use his tail much defensively when lifted out of his vivarium. Even so, he used it sufficiently to suggest how a fully adult crocodile might react. Moreover, when taken out for exercise and placed on the ground he would use a similar action to get away to a speedy start. He would press his tail to the ground to give himself the initial forward movement, the action being similar to the one he would use in water to start swimming. It was, however, an action in miniature, so to speak, as compared with anything a fully grown crocodile might do.

That was the real value of having Perce. His presence gave one the stimulus to learn more about the *Crocodilia* as a whole, to read the accounts of the latest observations and researches and to compare what was known with what one could observe in him at first hand. It may have been like studying mankind by observing the behaviour of an infant toddler but even that is not without profit. Apart from that, and compared with the more active birds and mammals that formed the bulk of our menagerie, Perce lacked any hint of vivacity. He was also totally lacking in aesthetic appeal; but then, he belonged to a stock that came into being when there were no human beings in existence to show aesthetic appreciation. This may be basically a facetious remark but it is not without virtue. Those things in animate nature which appeal to our senses so strongly today – the colours, the sounds and the graceful movements of animals – had to a large degree not yet emerged. They belong to the later stages of evolutionary history, to the geological eras that succeeded the Age of Reptiles when the rate of evolution speeded up enormously. Perce represented a page in earth's history and watching him enabled one, in imagination, to go back in time.

Crocodilians, including the crocodiles, alligators, caimans and gharials, are the oldest surviving reptiles, apart from the tuatara of New Zealand and the tortoises and turtles. These last two are highly specialized, with their heavy armour, and therefore less representative of the Age of Reptiles generally. Lizards and snakes, which make up the bulk of reptiles today, are more recent products from the speeding up of the evolutionary process, hence their colours, and greater speed of movement – even to the point of vivacity at times. They contribute little to the world of sounds, however.

The main ambition of crocodilians, judging from our experience with Perce, is to do a masterly nothing, except feed and rest. This is why these reptiles live so long and why as a race they have survived for such a long time. For most of their time they avoid strenuous effort and seek to conserve energy, even in the way they react to possible danger. In the face of a potential threat to their security they give forth a hissing growl, without making overt movements. It is a preliminary warning and if this is ignored the volume of the growl is increased. The next step may be to snap with the jaws, but a more likely move is to lash the tail. In this, the back is arched, foreshortening the body so that when the tail is whipped round its tip reaches to the tip of the snout. This puts the stout muscular tail between a would-be aggressor and the vulnerable flank within which lie vital organs, relatively unprotected; the only other vital internal organ is the brain, protected by its stout brain-case.

Perce was relatively easy to feed. Baby crocodiles eat insects. Then they graduate to fish and frogs and, at a later stage, to flesh and carrion. When first hatched the Nile crocodile is hardly a foot long and it increases in length by about a foot a year until it is eight or nine feet long, when growth in length is much reduced. The usual maximum length is twelve feet but Nile crocodiles of sixteen feet have been recorded and it is believed that, left unmolested, even greater lengths might be reached.

Insects are not difficult to obtain but those of the size likely to tempt a two-year-old crocodile sometimes presented problems. As a rule, we fed him with pieces of raw meat but we also gave him grasshoppers. These seemed to be attractive to Perce and it was of interest to watch his method of catching them. A grasshopper thrown on the water would struggle and soon reach the water's edge. Once its feet touched solid ground it would rest for a few moments. Perce would watch the grasshopper's progress, completely submerged except for his nostrils and eyes, which just broke the surface. Then he would move stealthily forward, rising slightly in the water as he did so. He seemed to be assessing the distance, ready for the final pounce. Because the eyes are on the sides of the head, a crocodile must take up position carefully, then make a sideways snap at its prey. While pausing to assess the distance the body is kept on an even keel by the hindlegs hanging down obliquely on either side. The tail meanwhile is waved gently from side-to-side in preparation for the final quick thrust that drives the animal forward to bring the snout level with the prey for the final sideways snap.

The whole action was reminiscent of a lizard stalking its prey on land. There is the same stealthy approach, the careful measuring of distance and the movement of the tail from side to side. In a lizard this waving of the tail registers the intention to move suddenly forward. Whether the tail is actually used in a thrust against the ground to give an initial impetus is an open question. The probability is that it is an intention movement pure and simple and there is the chance that it is no more than a relic from an aquatic ancestor.

The sequence of these actions, in the aquatic crocodilian and the terrestrial lizard, recalls the actions of a cat preparing itself to pounce. There is in a cat the stealthy approach, the assessment of distance, the slight movements of the paws as the cat puts them in position for the spring and the gentle lashing of the tail from one side to another. This last has never been satisfactorily explained.

One suggestion is that the movement of the tail catches the eye of the quarry and distracts it from noticing the other end of the cat with its signs of immediate attack. The weakness of such a theory is self-evident and if that is all that can be said in explanation then it seems not unreasonable to suggest that it is a relic from a far-off reptilian ancestor. We are at last beginning to appreciate that certain tricks of behaviour we find impossible to explain otherwise may be relics from the past.

We may go even further. A cat will also lash its tail at other times. We say it does so in anger; yet we deny, if we are purist in these matters, that animals are capable of the emotion which in ourselves we call 'anger'; but if they do not experience anger, they exhibit at times visible symptoms that are unconscionably like it. The greater likelihood is, however, that when a cat lashes its tail when disturbed this could be nearer the defence mechanism of the crocodilian, and its motivation comparable to the defensive whiplash of a crocodile when disturbed on land. Young crocodiles especially, and Perce made us aware of this, are prone to react in this way whenever they are upset or uncomfortable.

Indeed, very young crocodiles seem to make more use of the tail at all times than full-grown individuals. One of Maberly's more striking observations was concerned with a young crocodile's method of catching mosquito larvae, with its body curved and lashing the larvae into its mouth with its tail. Regrettably I did not learn of this until after we had parted with Perce or we might have given him the opportunity to demonstrate this, although at his mature two-feet length, the length to which he had grown while he was with us, such small prey might have been beneath his dignity.

Although Perce was only a youngster it was a wise precaution when handling him to hold him with your fingers round his throat or just behind his forelegs. This prevented his using his strong jaws and sharp teeth to protest. He seemed to resent this and responded by opening his mouth and hissing loudly. At the same time curious folds of skin appeared on his throat, the exact significance of which was not apparent.

There was, however, another display of resentment which he never did indulge with us, although we were hoping for it, because this might have shed more certain light on an ancient legend. The writers of classical times were in no doubt that crocodiles wept. They seem, however, to have been in two minds whether a crocodile sheds tears to lull intended victims into a false sense of

security or in hypocritical grief over the victim it has just consumed. They were all agreed, nevertheless, that crocodiles do shed
tears and we still speak of 'crocodile's tears' in allusion to an
insincere expression of sorrow. At the same time nobody really
believes that these poker-faced unemotional reptiles are capable of
feeling sorrow for any reason whatever.

In 1960, Jane was photographing a Mississippi alligator in
Paignton Zoo, using an electronic flash, which seems to have disturbed the reptile, for it went into an aggressive display. It hissed
violently and blew bubbles out of its eye-sockets, according to
Jane. Tears consist of liquid only. Bubbles consist of a gas
bounded by liquid but when the bubbles burst the gas escapes
and only the liquid is left, which would be sufficient to give the
impression of tears. As so often happens in photography the picture of a lifetime is lost through some trivial accident. In this
instance the frame for which the flash was used was the last on the
film and by the time the camera had been re-loaded the alligator
had submerged. Although Jane waited as long as she could it did
not surface to give a repeat performance.

Whether it was the electronic flash that disturbed the reptile or
some other circumstance, and whether the emission of the bubbles
was fortuitous or not, is beside the point. If this can happen with
one member of the *Crocodilia* it can happen with others – including
the Nile crocodile which ancient writers knew best. In the foundation of an erroneous belief or a legend, it is sufficient for a
phenomenon to happen rarely, or to be observed once only, for a
story to gain currency.

It could well be that whoever first observed what he took to be
tears issuing from a crocodile's eyes in those ancient times was also
struck by the contrast between them and the more or less permanent grin on the animal's face. Hence, in his opinion, the hypoc-
risy. The crocodile's grin has since been described as self-satisfied
and toothy; the second of these recalls that the external differences
between a crocodile and an alligator are very small. The only
infallible difference lies in the fourth tooth from the front in the
lower jaw on either side. In the alligator this fourth tooth fits into
a socket in the upper jaw and cannot be seen once the mouth is
closed. In the crocodile the fourth tooth fits into a notch in the
upper jaw and is still visible when the jaws are shut, giving the
characteristic toothy grin.

In any event, a crocodile has no need to express emotions or
mood in its face. Changes of mood, irritability or frustration or

any other of the factors that go to make up what is generally known as bad temper are expressed in its tail. Perhaps that is why we think of them as being as devoid of feelings as an animated steel gin-trap and seldom credit them with any versatility of mood. Yet it has been noted that crocodiles in one neighbourhood may be harmless to man and his stock, and those of the same species but in another locality dangerous to both. No fully adequate explanation of this has been put forward, but when we look more closely we find that, in spite of their evident adaptation to an aquatic life, there are many things that can try their patience. It may even be that the dangerous nature of some crocodiles can be traced to this.

In the course of an investigation carried out in the early 1960s it was found that, in clear water, the Nile crocodile makes no attempt to submerge, as if realizing it would be no better concealed if it were to do so. By contrast, in muddy water it quickly submerges when disturbed and remains underwater. From this it is possible to theorize that domestic animals and human beings are seized where the water is muddy because the crocodiles are hidden and so their prospective victims come upon them unawares, and therefore disturb their equanimity. It was also found that crocodiles become more and more aggressive as the level of the water drops until, when the pool is dry, even small crocodiles are dangerous to handle. When unable to submerge farther, because of the falling level of water, they grow uneasy, are irritable and much more inclined to hiss, snap or lash with the tail.

We might have learnt more by direct observation about these lethargic reptiles except that the time came when Jane was going abroad. As I have said, when these moments arrived we thinned out the ranks of our zoo. It was decided, not without reluctance, to find a new home for Perce. Jane remembered the lady at Worplesdon – and she readily agreed to take Perce into her care.

The story we had heard was basically correct. She did have a number of crocodilians, mainly alligators, but she did not keep them in her bath. She had a number of iron baths in a large shed at the back of the house, each filled with water heated by paraffin lamps. The whole arrangement was primitive but it seemed to have worked, for her favourite alligator, Daisy, lived for at least thirty-two years and grew to eight feet long.

Mrs Roberts was a remarkable person herself, flamboyant and theatrical. I found it was true that she took Daisy to the pond on the common and there the children played with her. Besides the iron baths and the pond on the common there was a pond in the garden of her house. A railway embankment ran nearby and the express trains from London to Portsmouth passed regularly each day. The driver and guard on the train, as well as regular passengers, would look down on warm sunny days to catch a glimpse of the alligators in the pond in the garden below, as their train rushed by at 70 m.p.h.

Mrs Guinevere Flora Elizabeth Roberts, to give her name in full, lived in a house called Weycliffe Cottage. The house nestled behind high hedges and the garden surrounding it was ill-tended, its unkempt shrubs and bushes seeming to conspire with its fruit trees and tall hedges to preserve the secrets within. The general absence of daylight within the house was unrelieved by the dim paraffin lamps that were lighted at nightfall. In the general gloom one could discern that the walls of the living rooms were decorated with African shields and spears and other exotic bric-à-brac as well as occasional animal skins. Tiger skins on the floor and a stuffed crocodile on the settee were in keeping with the rest of the décor. In this unusual setting, Miss Roberts, as she preferred to be called, in keeping with her vaudeville connections, or Katrina, her stage name, which she preferred even more, moved briskly and energetically. Her black, spangle-covered dress, reminiscent of a circus lady, and her slightly swarthy, almost Romany, complexion, made her even less well-defined in the general gloom, her flamboyance contrasting sharply with her female companion who hovered, withdrawn, reserved and unobtrusive in the background. So, between them, the two ladies contrived to focus a visitor's attention on the dominant inmates of the house. Chief among them was Daisy, who had been the star of a film *An Alligator called Daisy*.

This female alligator was the one most often taken out onto the common, guided by a lead attached to her left leg. Daisy could be lifted out of the water by a strong boy and would suffer any child to lie beside her with an arm over her shoulder. Those who knew this ménage well tell of seeing, not infrequently, one boy lying like this or even two, one each side of the alligator, with an arm over her shoulders. Nothing seemed to upset the even temper of the film star alligator.

Other alligators or crocodiles a visitor might encounter were William, who chain-smoked, Nellie Wallace, named after the famous comedienne, Moses and Andy Pandy. The last two were mere youngsters beside the venerable Daisy and apt to snap at one's fingers. But neither of them was so ill-tempered as George. Mrs Roberts always warned visitors to keep clear of him, a warning that often brought the reply: 'Don't worry, I shall keep clear of all of them.' There was a similar reaction to another inhabitant of Weycliffe Cottage, a nineteen-foot boa constrictor named Susie; and the two geese that were kept as 'guard dogs' were also to be treated with respect.

Looking at this set-up impartially it seemed slightly superfluous to have had anything in the nature of watchdogs to warn off intending housebreakers. Yet having geese to do the job reflected another of Mrs Roberts' characteristics. Although she loved her crocodilians as other people love their dogs, she made no secret that she thought canines were 'filthy creatures'. She had no time whatever for them.

William was not the only one that smoked. Mrs Roberts was always ready to exhibit her charges at fund-raising fairs and fêtes, and expecially events aimed at raising money for the Brookwood Hospital. The crocodilians were transported by those friends and acquaintances possessing lorries. There are photographs in existence showing one or other of the alligators with a smouldering cigarette, cigar or pipe between its teeth, possibly also with a paper hat on its head (or, rather, on its neck, for there is little left of a crocodilian's head behind the eyes). Apparently one of Daisy's accomplishments was to smile to order. Mrs Roberts would say 'Smile, Daisy', and this most accommodating reptile would obligingly open her mouth.

Anyone that can teach a crocodilian of any kind even a simple trick such as smiling to order deserves full marks!

I mentioned a little earlier in this chapter the supposed differences in temperament in crocodiles of different localities and of various suggested explanations for this. On the showing of Mrs Roberts' pets it could be that there is a very simple reason: that the crocodiles of, say, the Blue Nile are inoffensive and those of the White Nile are aggressive simply because the former, like Daisy, are amiable, and the latter, like George, are testy. After all, crocodiles tend to form groups – and these isolated populations in the big rivers could lead to discrete gene-pools!

This, then, was the home to which Perce was taken and, notwithstanding what has been written here, we had no qualms about his well-being. The last we heard of him he had grown to a length of six feet – from a purse to a handbag. Reports spoke of his being well and lively, and the reports included some from independent eye-witnesses.

Then Mrs Roberts went to live in Cornwall and even those in Worplesdon who were in any sense her close acquaintances lost track of her. She died in the early 1970s and, despite the ease of communication generally in this modern age, it has been imposs- ible to find out what happened to her reptilian charges after her death. We can but hope that they, Perce among them, found haven in one or other of the better zoos in the south-west corner of England.

17 *Reverie*

The characters that form the subject of this book are the 'stars', the constellations that stand out prominently against the background of lesser lights as seen in a cloudless night sky. These lesser lights were not without their interest and with the stars they build up a background of zoological experience which, in the evening of my life, I can sit and look back on with pleasure and deep satisfaction.

At such moments I think of the many thrushes and blackbirds, the greenfinches, the bullfinches, the starlings, nuthatches, woodpeckers, and even the occasional house sparrow, that passed through our hands, coming to us either as adults that had been injured or as fledglings that needed to be hand-fed. There were the shrews – the common, the pigmy and the water shrews – the several moles that we had as pets, the fieldmice, the yellow-necked mice, the field voles and other small rodents, the grey squirrels and the red squirrels. They all had their charm, and all gave us further insights into the lives of insignificant members of our wildlife.

As I cast my mind back over the past twenty-five years, counting the many species that came into our care, I see the polecats, the mink, and the Scottish wild cat, not forgetting the raccoon. I had hoped it would demonstrate to me how this animal can untie knots, which writers on raccoons delight in mentioning. Unfortunately, the 'coon is nocturnal and gave us little in return for its board and lodging. I think of the several kinds of mongooses that we had, of the marmoset we gave hospitality to for a while, and the bush-babies that we took over from people who had bought them and tired of them.

Among the medium-sized or larger birds I recall the barn owl and the little owl, the several magpies, the jackdaws and the ravens, the kestrels and the sparrow hawks that we rescued. There was the cuckoo that I had hoped to keep indefinitely to see whether it would hibernate, as several writers have suggested it will, but which thwarted this ambition by bashing itself against the wire-netting on the south side of its aviary when the time for migration arrived, so that we were forced to give it its liberty for fear it might damage itself fatally. We even had a goat for a while which played in a delightful way with Jason, our boxer-cross, but went out of favour slightly when it pulled its tethering iron one night and walked across the kitchen garden sampling everything in its path. It cut a swathe through the fruit trees, the fruit bushes, the brassicas and the rootcrops, leaving behind it a trail of destruction such as one might associate with a miniature tornado.

Our acquisition of zoological knowledge was not wholly confined to those animals we deliberately imported. Some of the more surprising discoveries came from the hangers-on in the aviaries and pens. An example that springs to mind is of the earthworms that used to congregate under the bowls used to give our birds their bath water. The ground there was always damp and there was one large bowl which we sunk slightly into the earth. Every day it had to be lifted out to be emptied, cleaned and refilled with fresh water. Inevitably much of the rejected water drained down into the depression in which the bowl normally rested, and where the earthworms congregated. One day as I lifted the bowl out I had a fleeting glimpse of an earthworm coiled as tight as any spring. Before my very eyes it uncoiled – so that it leapt a full six inches into the air to land on the surface of the ground. In retrospect, I sometimes wonder whether I actually saw this and yet I have the note that I made at the time, because it was so incredible.

The harvest mouse is perhaps one of the most insignificant of mammals, with regard to its size, but it is one of the most charming. We were presented with four of them and they lived in a glass cage decorated with herbage, including stalks and ears of wheat, the whole standing on a table in our sitting-room. There was no offensive odour from them, no noise and only charming antics to delight the eye.

During the years that we had them, the harvest mice in their crystal cage were a showpiece for visitors and it was not long before it began to dawn on us that they were not always visible. For example, they seemed never to show themselves at our teatime, so that anyone that came to tea was denied the sight of these diminutive rodents. One Sunday it was raining when we got up and it poured with rain throughout the rest of the day. I decided this was a suitable day to carry out more extensive observations on the timing of the activities of the harvest mice. Immediately after breakfast I settled myself in an armchair beside their cage with a book put in such a position that any slight movement in the cage would impinge itself on my peripheral vision (that is, I would see it out of the corner of my eye). I also had a notebook, a pencil and a clock. So from early morning until nearly midnight I noted the times when the harvest mice were out and about and when they disappeared into the thick mat of chewed up grass fragments on the floor of the cage in which they had their sleeping quarters. I found that throughout the day they alternated three hours activity and three hours rest.

Unbeknown to me until the following morning Richard, then on holiday from school, sat up throughout the following night to continue what I had done during the day. The harvest mice repeated their behaviour: three hours on and three hours off.

Up to this time it had always been assumed that harvest mice were diurnal, active throughout the hours of daylight, and that they rested throughout the hours of darkness. What I had established, with Richard's aid, was that every twenty-four hours in their lifetime was split into three-hourly periods, more or less, of alternating activity (feeding, climbing up and down the wheat stems) and resting.

Two days later I mentioned this to Peter Crowcroft, then Curator of Mammals at the Natural History Museum, and he showed no surprise. He told me he had just read an account in a scientific journal that showed that house mice had three hours on and three hours off, whereas previously it had always been assumed that house mice were nocturnal. This pattern was part of what has since been called the circadian rhythm. I had made a new discovery only to find that somebody else had found it in the house mouse and got in before me. I regretted that I did not have the wisdom to write up an account of what I had seen in harvest mice and publish it in a scientific journal.

Another incident which comes back to mind when I sit and contemplate this past quarter of a century is of the roe deer we have tried to help, and more especially of the roe deer kid that a woodman brought me one day, saying it had been abandoned by its mother. We bottle-fed the kid and later took it into the woods at a point where the woodman said he had found it. We set it down on the ground, withdrew and waited in the hope of seeing the mother searching for it and finally finding it. Naturally, Jane was equipped with a camera at the ready to photograph the happy reunion if it took place, although our primary motive was strictly humanitarian. For a while the roe kid lay crouched on the ground with head erect watching all around it. Then, as if impatient, it rose to its feet and uttered a most heart-rendingly piteous call. One could almost hear in the tones of that call a human child desperately and piteously calling 'mamma'. The call was also loud for so frail and small an animal, and although it was repeated no roe doe put in an appearance and we were forced to take the kid home again and continue its bottle-feeding.

The many wild birds that passed through our hands included swifts, swallows, house martins, grebes, pigeons and a swan that needed care and attention. The most surprising were the bitterns. We rescued one starved and feeble, in the depths of a hard winter, and another under the same circumstances a few years later. The paramount lesson I learned from these two incidents was the remarkable camouflage effect of a bittern's plumage. This is mainly brown of various shades, broken and dappled in such a way that one could see how easily it could blend into the background of dead leaves, reeds or other vegetation – as was demonstrated whenever we placed it in such an environment. What was the more surprising was that if the bittern was placed on the ground in the middle of a field, it did not show up against the bright green of the grass unless you had your eye on it. For example, if you took your eye off the bittern and looked around and then looked down to see where it was you found it necessary to search for it, so inconspicuous was it against the bright green background.

The time came when the Ministry of Education in its wisdom (or otherwise) closed the village school because they regarded it as an uneconomic unit, instead of seeing it as a vital element in the social structure of village life. Everyone here was very sad about it. Then the school and its schoolhouse with the surrounding land that had done duty as a playground were put up for sale. Jane and Kim inspected it and decided that this was precisely what they wanted. The idea was that the school would make a first-class photographic studio, the schoolhouse could be used for living accommodation, and the extensive playground was just what Jane needed to fill with animals.

We had been privileged to have Jane and her husband and her two children living in our house for a number of years. The time had come when they quite naturally spread their wings and went to live a stone's throw from us. It was natural, too, that all the aviaries and pens in the garden of Weston House should be transferred to the playground in front of their new house where, as the months passed, and as Jane's propensity for filling every corner with an animal, had free rein, the playground became an 'overloaded ark'.

A few aviaries were left here against the possibility of birds being brought to me requiring temporary accommodation, but otherwise the 'zoo in my garden' has ceased to exist. However, it has taken on a new lease of life in and around what was formerly the village school and is now the most commodious studio.

The garden at Weston House has now reverted to its former state of being one of the most beautiful of gardens, devoted to trees, shrubberies and flower-beds. It has lost nothing in charm, but there is less eye-catching activity, as far as the casual visitor is concerned. As I walk about the garden I recall that here, on this spot, stood the owls' aviary, here stood the foxes' pen, and the spirits of these animals on whom we lavished so much labour and affection seem to be present. This is purely subjective and cannot be in the mind of any person other than those who knew our zoo in its hey-day. For me, as I walk around the countryside, I carry this still further. If a jay flies up at my approach and disappears rapidly towards the nearest woods I watch it go and think of it as just another Jasper. Any rook that flies overhead is in my terminology a Corbie, and I am disappointed when I *tch-tch* to an owl perched in a tree in broad daylight and it does not respond as our tawny owls used to.

Index